Make: BOOKS
meet the author

Make: Fun!
author Bob Knetzger wants your kid to drink zombie barf juice
Written by Gretchen Giles

Make: Fun!
BOB KNETZGER
40 Fun Projects You Can Build!

Make & Make

Create Your Own Toys, Games, and Amusements

Flip-Book Inside! Try It!

BOB KNETZGER DIDN'T INTEND TO SPEND HIS CAREER AS A TOY DESIGNER, but that's perhaps because he didn't reckon on the positive power of Big Jim. Mattel's answer to the GI Joe doll, Big Jim was called into service as a cheap figurine in a toy prototype Knetzger built in college. When a Mattel recruiter visited campus, Knetzger's use of Big Jim caught his eye, and Knetzger soon found himself employed in the toymaker's preliminary design department. He was in lavishly creative company.

"**There were magicians and mathematicians, chemists, seamstresses. Just about any skill that could be pressed into service coming up with toy designs, was,**" Knetzger told *Make:*. "**There were 90 or so creative people. It was an in-house think tank.**" Knetzger stayed at Mattel for seven years, using his degree in industrial design to create electronic toys and helping to produce one of the first dedicated mini computers for kids.

He eventually left to co-found Neotoy with partner Rick Gurolnick. The company produces everything from STEM toys to the free "gifts" found in Cap'n Crunch cereal, but perhaps their most enduring invention is the Doctor Dreadful line of gruesome candies.

"**Looks gross, tastes great!**" Knetzger said, cheerfully recited the tagline. "**It's a line of mad scientist activity toys where you make strange concoctions and everything you make you can eat or drink.**" With the popular Zombie Lab set, kids make what Knetzger offhandedly refers to as a "zombie barf drink." With Bubbling Brains, they inject a candied potion into a molded skull and devour the gelatinous worms that result. "**With the proper toy components, you can make something that's really gross but pretty tasty,**" he said.

As it turns out, industrial design makes a perfect background for gross. "**It's like being an architect of products,**" he explained. "**You'll often work on something you've never worked on before. That outsider's viewpoint is important. Being fresh about something and not knowing ahead of time what will or won't work allows you to blindly stumble ahead and find something that does work.**"

Even if that's a nice big glass of zombie barf juice.

Find *Make: Fun!* at Makershed.com and other booksellers

CONTENTS

SPECIAL SECTION

BIOZONE

26

ON THE COVER: Exploring science with an eerie blend of ferrofluid and glow stick solution. Photo: Alan Rockefeller.

Hep Svadja

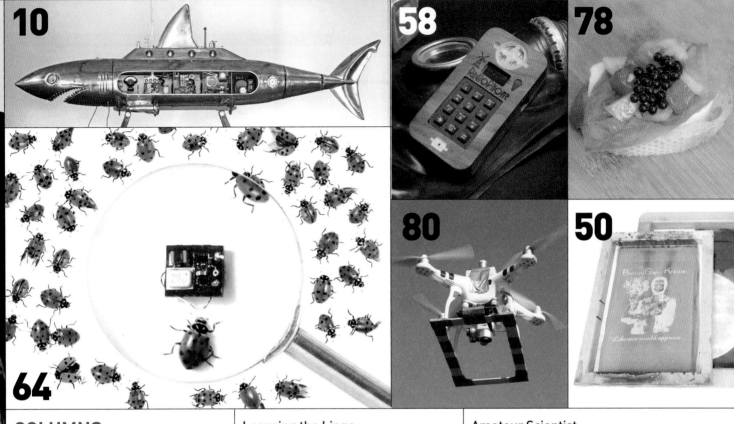

Christopher Potter, Hep Svadja, Forrest M. Mims III, Tim Deagan

Make:

EXECUTIVE CHAIRMAN & CEO
Dale Dougherty
dale@makermedia.com

CFO & PUBLISHER
Todd Sotkiewicz
todd@makermedia.com

VICE PRESIDENT
Sherry Huss
sherry@makermedia.com

EDITORIAL

EXECUTIVE EDITOR
Mike Senese
mike@makermedia.com

PROJECTS EDITOR
Keith Hammond
khammond@makermedia.com

SENIOR EDITOR
Caleb Kraft
caleb@makermedia.com

MANAGING EDITOR, DIGITAL
Sophia Smith

PRODUCTION MANAGER
Craig Couden

COPY EDITOR
Laurie Barton

CONTRIBUTING EDITORS
William Gurstelle
Charles Platt
Matt Stultz

CONTRIBUTING WRITERS
Bonnie Barrilleaux, Donald Bell, Brian Berletic, Gareth Branwyn, Jordan Bunker, Bill Chellberg, Emily Coker, Larry Cotton, Patrik D'haeseleer, Tim Deagan, DC Denison, Stuart Deutsch, DIYbio, Paloma Fautley, Jose Gomez-Marquez, Bob Knetzger, Quitterie Largeteau, Lisa Martin, Michael Martin, Forrest M. Mims III, Bob Murphy, Katherine Ozawa, John Edgar Park, ProgressTH, Jendai E. Robinson, Rick Schertle, Tom Schneider, Nicole Smith, Michael Weinberg

DESIGN, PHOTOGRAPHY & VIDEO

ART DIRECTOR
Juliann Brown

PHOTO EDITOR
Hep Svadja

SENIOR VIDEO PRODUCER
Tyler Winegarner

LAB/PHOTO INTERN
Sydney Palmer

MAKEZINE.COM

WEB/PRODUCT DEVELOPMENT
David Beauchamp
Rich Haynie
Bill Olson
Kate Rowe
Sarah Struck
Clair Whitmer
Alicia Williams

CONTRIBUTING ARTISTS
James Burke, Monique Convertito, Colin Johnson, Bob Knetzger, Andrew J. Nilsen

ONLINE CONTRIBUTORS
Lance Akiyama, Cabe Atwell, Duane Benson, Erich Campbell, Kathy Ceceri, Jon Christian, Jeremy Cook, Josh Elijah, Chris Fox, Lauren Glaubach, Mike Hinkle, Shawn Jolicoeur, Art Krumsee, Goli Mohamadi, Larry Moss, Andrew Salamone, Thomas Sullivan, Andrew Terranova, Glen Whitney

SALES & ADVERTISING
makermedia.com/contact-sales or sales@makezine.com

SENIOR SALES MANAGER
Katie D. Kunde

SALES MANAGERS
Cecily Benzon
Brigitte Mullin

STRATEGIC PARTNERSHIPS
Allison Davis

CLIENT SERVICES MANAGER
Mara Lincoln

BOOKS

PUBLISHER
Roger Stewart

EDITOR
Patrick Di Justo

MAKER FAIRE

PRODUCER
Louise Glasgow

PROGRAM DIRECTOR
Sabrina Merlo

MARKETING & PR
Bridgette Vanderlaan

SPONSOR RELATIONS MANAGER
Miranda Mota
miranda@makermedia.com

COMMERCE

SENIOR PRODUCT DEVELOPMENT
Audrey Donaldson

PRODUCTION AND LOGISTICS MANAGER
Rob Bullington

PUBLISHED BY
MAKER MEDIA, INC.
Dale Dougherty

Copyright © 2017
Maker Media, Inc.
All rights reserved.
Reproduction without permission is prohibited.
Printed in the USA by Schumann Printers, Inc.

Comments may be sent to:
editor@makezine.com

Visit us online:
makezine.com

Follow us:
@make @makerfaire @makershed
google.com/+make
makemagazine
makemagazine
makemagazine
twitch.tv/make
makemagazine

Manage your account online, including change of address:
makezine.com/account
866-289-8847 toll-free in U.S. and Canada
818-487-2037,
5 a.m.–5 p.m., PST
cs@readerservices.makezine.com

CONTRIBUTORS

What's the next skill on your list to learn and why?

Jendai E. Robinson
Mountain View, CA
(Microbial Fuel Cell)
I would like to learn more computer coding skills because it's becoming increasingly important to have strong computer skills in today's job market.

Bonnie Barrilleaux
San Francisco, CA
(Double Helix Daiquiris)
Inspired by seeing Adam Savage create a custom plastic coupling for a ping-pong machine gun, I want to master the lathe so I can make custom parts for my projects.

Tom Schneider
Montreal, PQ, Canada
(Teeny-Tiny Spy Bug)
I plan to work with pulse power applications, e.g., coilguns and railguns. I'd like to improve my understanding of physics, pulse power electronics, and learn about the possibilities and constraints.

Colin Johnson
St. Paul, MN
(BioZone illustration)
The next skill(s) on my list would be to learn a few new painting programs, and in keeping with the nature of *Make:* magazine, to experiment and make better images as a professional artist.

Sydney Palmer
Berkeley, CA
(Lab/Photo Intern)
I've currently been teaching myself how to carve and print linoleum. I really want to get the printing process down so I can make some finished pieces!

Issue No. 56, April /May 2017. *Make:* (ISSN 1556-2336) is published bimonthly by Maker Media, Inc. in the months of January, March, May, July, September, and November. Maker Media is located at 1160 Battery Street, Suite 125, San Francisco, CA 94111, 877-306-6253. SUBSCRIPTIONS: Send all subscription requests to *Make:*, P.O. Box 17046, North Hollywood, CA 91615-9588 or subscribe online at makezine.com/offer or via phone at (866) 289-8847 (U.S. and Canada); all other countries call (818) 487-2037. Subscriptions are available for $34.99 for 1 year (6 issues) in the United States; in Canada: $39.99 USD; all other countries: $50.09 USD. Periodicals Postage Paid at San Francisco, CA, and at additional mailing offices. POSTMASTER: Send address changes to *Make:*, P.O. Box 17046, North Hollywood, CA 91615-9588. Canada Post Publications Mail Agreement Number 41129568. CANADA POSTMASTER: Send address changes to: Maker Media, PO Box 456, Niagara Falls, ON L2E 6V2

PRINTED WITH SOY INK

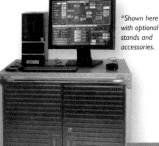

A DIY'd
DIY Magazine
& The Little
Boat That Did

KUDOS FOR A BRAVE LITTLE BOAT

The story of Damon McMillan's *SeaCharger* ("The Little Boat That Could," *Make:* Volume 55, page 14) was amazing. I hope he will build a *SeaCharger II* and try again to cross the Pacific. Perhaps this time from California to Japan, where their robot culture will certainly honor him and his creation with the glory they deserve!

–Kenneth Scharf, via the web

I have just finished reading ["The Little Boat that Could"]. It was outstanding! I crave more information, especially regarding some of the details of his "satellite modem"

JUST LIKE THE REAL THING

» Kids [Zhia and Ocean] bought a *Make:* magazine subscription for their dad, but the first issue wasn't going to arrive in time so they made him this.

–Robyn, Quadra Island, British Columbia, via Twitter

control and information channel, and his magnetically coupled main propulsion motor. A follow-on article, please!

Mr. McMillan might consider asking NOAA to issue a Notice to Mariners (NOTAM) giving *SeaCharger's* last known position, a request for passing vessels to be on the lookout for it, and "pick-up" procedures (e.g., How does one shut it off if recovered? Don't worry about hazards; it does not contain explosives, etc.).

–Bruce McCandless II, via email

Author Damon McMillan Responds

Thank you so much for your interest and kind words.

The satellite modem is called a "Rockblock," and it's made by a company in England. It's just a little palm-sized board that costs about $250. Every time it sends or receives a message, it's roughly 30 cents. I can send basic commands to the boat, for example to give it a new waypoint or to turn the motor on or off. It is connected to and controlled by an Arduino.

The magnetic coupling is actually pretty standard for underwater motors. It just has two concentric circles of magnets with a waterproof barrier in between. The inner circle is coupled to the motor and remains dry while the outer circle is coupled to the propeller and gets wet. The outer circle of magnets follows the inner circle because of the attraction between the alternating

» Brie Finegold takes a well-deserved breather after building the "One-Day Wood-Fired Pizza Oven" (*Make:* Volume 53, page 34) with her dad, Joe, in Santa Barbara, California.

magnet poles.

Incidentally, just yesterday [Jan. 11, 2017] the boat was picked up by a container ship on the way to New Zealand. So it'll arrive there tomorrow and perhaps spend a few months in a museum there.

If you go to the website seacharger. com, you can see more pictures and other info. ⊘

EDITOR: Damon also recognized the author of the second email as Bruce McCandless, former U.S. astronaut — thanks for being a reader, Bruce. You too, Kenneth, Brie, Joe, Robyn and family!

Subversive Science

BY MIKE SENESE, executive editor of *Make:* magazine

Hep Svadja

Community-driven innovation is the key aspect of the maker movement, both in the tools we use and the projects we build. We've seen it with the rise of accessible electronics prototyping boards and digital fabrication machines; We've seen it as challenging and expensive equipment has become compact, affordable, and simple to use by all; And we've seen it in the communities of enthusiasts who use these devices to constantly surprise us with interesting and useful projects. Community-driven innovation, after all, is the basis for this magazine, for Maker Faire, and for the energy of the movement overall.

Biohacking is on a similar journey of innovation driven by a dedicated community, with possibly greater potential for impact — and for this reason, we're highlighting it in this issue of *Make:*. Today it's easier than ever to do serious science, as (once again) new, accessible tools have allowed for home labs and biohacking makerspaces to set up in almost every region, letting amateurs pursue and collaborate on advanced project ideas once reserved for academics and professionals only.

As a result, we're seeing creative yet beneficial amateur projects coming from all over — concepts that have the potential to make huge impacts to the world, from saving lives to assisting with natural conservation. Two covered in this issue, OpenAPS and the Open Insulin project, promise to enhance the world for diabetics by giving them device control and medication alternatives. Meanwhile, biohacking group Pembient aims to end the destructive rhino horn trade by flooding the market with synthetic keratin rhino protein; some disagree on the tactic and its effectiveness, but the creativity of the concept shows the spirit of these groups. And the Real Vegan Cheese team is working on a similar process to create dairy products by sequencing milk proteins in yeast, rather than relying on cows. Their output would then be used to make animal-free cheese using traditional techniques. By controlling the sequencing, traits like lactose can be left out and allow people with dietary restrictions or intolerances to enjoy lab-made dairy alternatives.

The list of possibilities goes on and on, attracting and energizing new groups, bringing new users into biohacking spaces, and inspiring even more creativity. We're watching the enthusiasm for this space accelerate and are excited to see where it leads. Medicines for impoverished regions, printed organs and bones, a fuel revolution — it's all plausible and could very well come from you biohackers. Keep it up. ◉

Let us know what you're working on:
editor@makezine.com

MADE ON EARTH

Backyard builds from around the globe

Know a project that would be perfect for Made on Earth?
Let us know: *makezine.com/contribute*

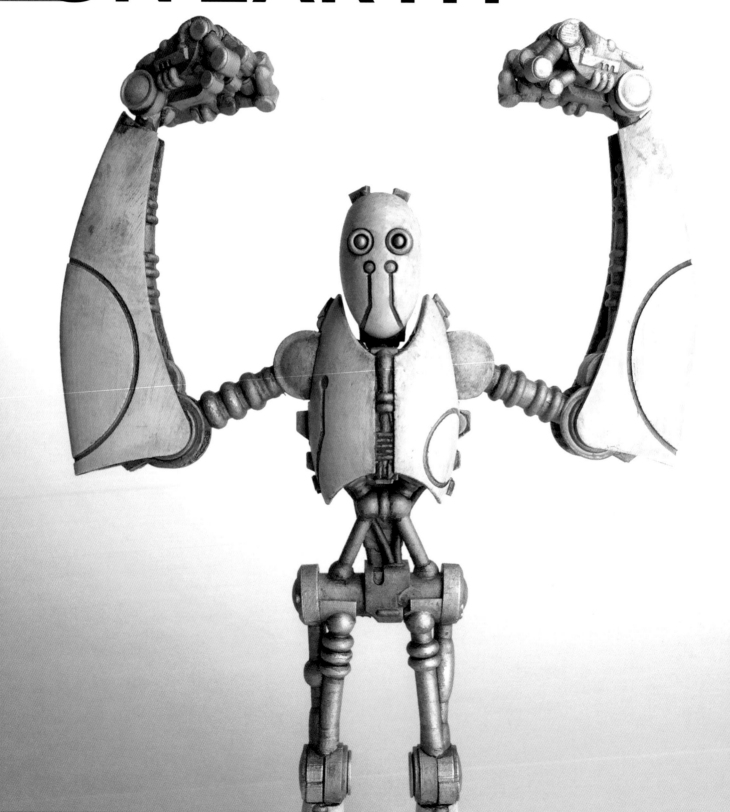

DECO DROIDS

R2DECODESIGN.COM

Brett Mich, like many of us, has an unexplainable love of robots. The Wisconsin-based designer is a toy inventor by trade, but spends his free time crafting one-of-a-kind robot sculptures. "Initially, designing robots was a way to add to my personal collection," he says. But after giving one away as a secret Santa gift at work, Mich was encouraged to open an online shop for his creations called R2Deco Design.

Each robot begins as a series of pencil sketches, and later reimagined on his computer using Rhino CAD software where he creates working joints and breaks up the design into distinct pieces to help with assembly. The parts are then brought into the real world with a 3D printer using PLA filament, and treated with an exhaustive process of painting and weathering to achieve a vintage robot realism.

"Like a sculptor, I make one version of a robot design and then I move on. It's more fulfilling to create a variety of robots than to make the same one over and over."

Mich takes his inspiration from sci-fi art books and everyday objects, often creating a series of designs around a theme, such as vintage cars or cartoons. For the final step, each sculpture is photographed and posed in multiple positions to showcase its personality before being placed on his website for sale.

"I feel this brings the robots to life and gives people a better sense of what they really are — posable art," Mich says.
—*Donald Bell*

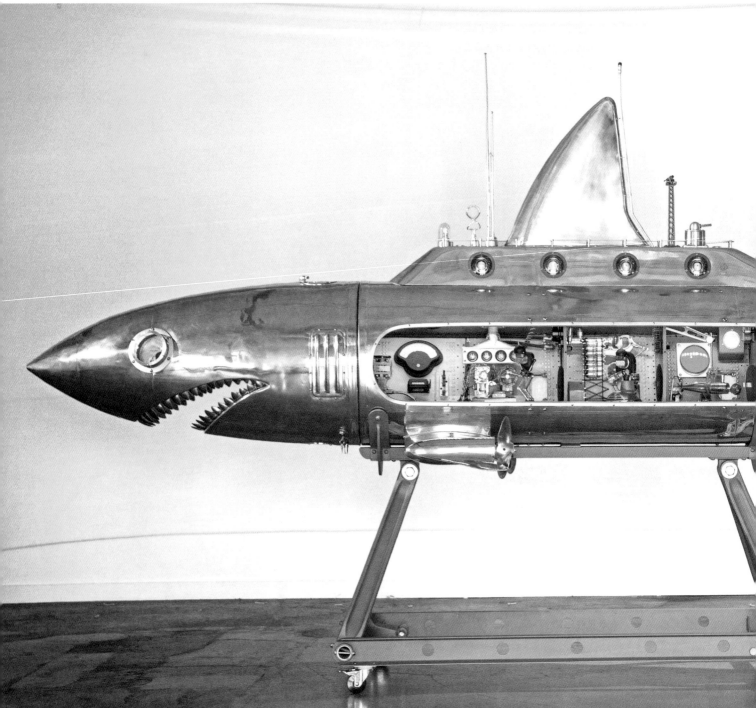

Nemo Gould

SUBLIME SUB

NEMOGOULD.COM

Christopher Potter

Fine artist and self-described master hoarder **Nemo Gould** conjures up fantastic sculptures made entirely of found objects. Rich wood and gleaming chrome catch the eye as they cycle through their kinetic loops, while tentacles and antennae extend in a playful fashion like a sci-fi comic book come to life.

The *Megalodon* is Gould's latest work, a 16-foot-long salvaged fuel tank from an F-94 bomber plane's wing. The shark has working propellers for fins, and a tail that glides back and forth ominously. A cutaway on the side reveals various boiler and control rooms, each with their own delicately installed moving parts. It's packed full of tiny human figures and whimsical creatures alike, all in mid-task as they operate their predatory underwater vessel.

The project took Gould a little over two years to finish. "I'd wanted to make a cutaway vessel for years, and had been putting objects aside for that purpose," he explains. "I know it sounds backwards, but the tank was the last missing piece." He found it at an aircraft salvage business, and from there he was able to assemble the final sculpture.

Gould says his process is a lot like solving a puzzle. "I maintain an extensive collection of things that I feel strongly about one way or another," he says. "The challenge is to find which of the million potential relationships between these things could lead to the best art." More so than his skills as an artist, machinist, fabricator, woodworker, et al., Gould says that "maintaining a vast, organized library of seemingly random objects is the real trick." —Sophia Smith

PAPEROPOLIS PAPERHOLM.COM

Back in August 2014, artist **Charles Young** took on a one-a-day, yearlong project creating kinetic paper sculptures, which he calls *Paperholm*. Young needed a new project that summer after completing his Master of Architecture degree at Edinburgh College of Art, and he wanted to tackle something that would "keep [him] making something every day." He had made little paper models before and decided it was something he could complete on a daily basis. Thus, his project began.

Each model is made using just 200gsm watercolor paper and PVA glue, but they come to life in gallery shows and as GIFs on his website, looping diligently as quaint amenities of some alternate papercraft reality. Young constructs the buildings first thing in the morning, and spends anywhere from 20 minutes to 3 hours on each, depending on complexity, although he's much faster now than when he started. "Once you get 100 or 200 days in you suddenly realize what a mass of

work you've produced without spending very much time on it each day," he says. While it was difficult to stay dedicated, the daily rhythm was hugely beneficial for his creativity and productivity.

Although he completed the original 365-piece collection in August 2015, he resumed in November of that year, and has been making more daily buildings ever since. His humble village continues to sprawl into a verifiable metropolis.

—*Nicole Smith and Sophia Smith*

Charles Young

METAL MIXOLOGIST

TWOBITCIRCUS.COM

Los Angeles-based entertainment makers **Two Bit Circus** have a mission of creating outlandish yet engaging projects, from high-tech rocking horse races to escape rooms to a dunk tank that envelopes the person inside with flames instead of water. For their annual Anti-Gala fundraiser this past November, they wanted a centerpiece worthy of their mission of fun, social, interactive technology. Their solution? Create a sleek robot bartender jukebox to make and serve tasty beverages for all the events' guests.

With just three weeks to work, the team set out to build their drink-slinging droid. They crafted the bartender's body using parts from surplus electronics shops around the LA environs, and animated it with powerful servos controlled by a Pololu Maestro board. They used a peristaltic system from Party Robotics to dispense alcohol and mixers into tumblers held by the robot, which it shakes and pours into a guest's glass through a spigot. A custom system run on a Raspberry Pi offers four beverage options through oversized pushbuttons, plus a fifth button to pour the drink.

The finished automaton looks like it was taken straight out of George Jetson's favorite happy hour lounge, largely in part to the mid-century encasement that houses it all. "We started off with a 'Zoltar' style cabinet, but decided that it wouldn't fit with the much more contemporary-looking robot," Two Bit Circus' industrial designer Chris Weisbart writes. "Through years of collecting books on industrial and product design, I had one on the history of jukeboxes. The 'aha!' happened when we flipped through the book and found the AMI H from 1957, an astoundingly beautiful design that we felt we could adapt to our needs."

The reception was a hit — "People went ape over the design," Weisbart says, adding "We had a hard time filling the dispensing bottles because folks wanted to talk nonstop about it and kept pressing buttons as we were trying to refill the machine for drinks. The party was phenomenal and the bartender was mobbed all night long." —*Mike Senese*

RETROFUTURE REMOTE MAKEZINE.COM/GO/STEAMPUNK-REMOTE

When you pick up the remote control to your stereo, are you pleased with how it looks? Most of us don't put much thought into it, because they're typically so bland that they're nearly invisible.

Michael Greensmith wanted to take a different approach. "I wanted a remote control that's fun to use and would look good next to my music system driven by an iPhone, housed in a 1940s mantle clock, which it operates," he says. "I just kept thinking of great sci-fi novels and comics I'd read as a kid."

His remote certainly looks the part of some kind of retrofuturistic science fiction prop. The wooden case and typewriter keys give it the heft and feel of a piece of old equipment and the eerie blue glow emanating from within keeps you wondering just where in time it came from.

To make it work, the typewriter keys extend down to poke at the guts of a Yamaha stereo remote nestled inside. Low tech for sure, but functional! A simple LED is attached to the remote batteries for that blue acrylic light effect. —*Caleb Kraft*

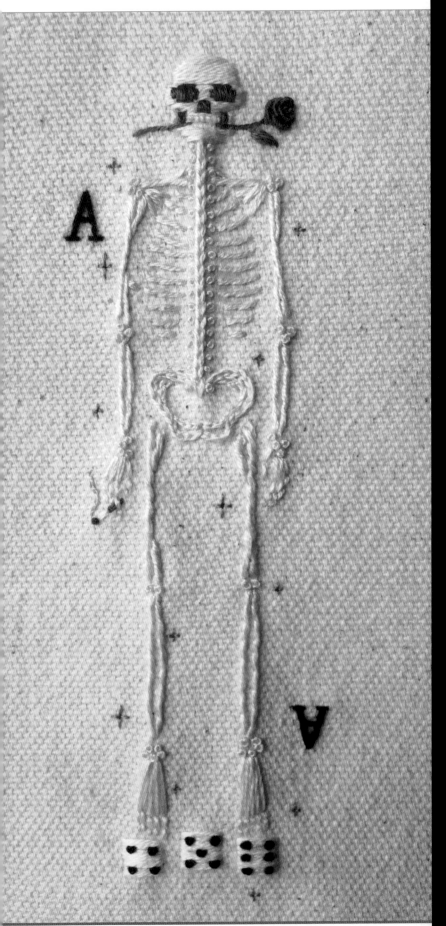

NIMBLE NEEDLEWORK

TINYCUPNEEDLEWORKS.COM

Artist **Britt Hutchinson**, better known as Tiny Cup Needleworks (@tinycup_), has only been doing hand embroidery for about three years, but she's already amassed over 74,000 Instagram followers with her charmingly teeny tiny needlework.

She often photographs her work alongside a quarter to highlight its miniscule scale. "There's something incredibly satisfying about meeting the challenge of executing a detailed piece inside of a small space," she says.

Despite its small size, her work seems to leap off the cloth at you. It's delicate and unassuming at first, but the longer you look, the more intentional detail you can see in the stylistic variety of her stitches and knots. Techniques like bullion knots and French knots add texture and three-dimensional shape to her images. In fact, her gravitation toward skeletons was at least partially inspired by the resemblance that a line of French knot stitches has to a spine. "It was in those shapes that I started to piece together images, sort of like a puzzle," she explains. "Eventually my process became an amalgamation of manipulating traditional stitches and puzzle piecing all of those shapes into a composition."

While flowers, candles, books, and moons make regular appearances in her romantic, mysterious motifs, her pieces primarily feature skulls and skeletons. "My work is based on human emotion, and all people are just bones at the core of them," says Hutchinson. "Anyone has the opportunity to project their own emotions onto my imagery, and possibly find their own catharsis in seeing what they feel projected back at them. It's like when you hear a song that 'gets you,' and you feel less alone. I find a great deal of comfort and solidarity in those connections."

—*Sophia Smith*

Use these tips and tricks to find success — as a creator and a backer alike

CROWDFUNDING CHEATSHEET

Written by
Gareth Branwyn

Illustrated by
Andrew J. Nilsen

CROWDFUNDING IS A FAST-MOVING TARGET THESE DAYS, WITH NICHE SITES EMERGING, a growing ecology of support services, the rise of equity funding, and more. After talking to makers who've both successfully launched and backed projects, and to those in the crowdfunding business, we put together this collection of tips, tricks, and helpful resources for makers considering launching or backing projects.

TIPS FOR PROJECT CREATORS

Here are a few hard-won, often overlooked lessons in launching a healthy campaign, as told to us by successful project creators.

LEARN THE SPACE BEFORE YOU LAUNCH

Do research, lots of it, before launching a campaign. Look at all of the platforms, compare strengths and weaknesses, talk to those who've launched on those sites. Back a number of projects. Bookmark campaigns that appeal to you, how they're being executed, and think about whether you can emulate them.

KEEP VIDEOS SHORT AND TO THE POINT

Doc Popular, a musician and comic book author who's created six successful campaigns and backed over 380 others, says: "Keep your videos short." This was a sentiment we heard from many other successful project creators. Videos really are an ad — important for most campaigns — but creators get carried away. A three-minute video is fine for most projects, and when you look at the viewing stats on the back end of sites like Kickstarter, don't be surprised to find that the majority of people only watch the first minute or two.

MAKE UPDATES FUN AND INFORMATIVE

Many project creators take pride in keeping their backers updated during the entire funding process and beyond. "I want them to feel like they're part of the process, so I share progress pics and ramblings as often as I can," Doc Popular says. "I may even try to set calendar reminders to share some bits of news way after a project has ended. If I make a mistake along the way, I like to share it early and honestly with my backers. There may be some complaints, but it's better if I deal with it early on rather than just avoid talking about it until things get out of hand." I did my own successful Kickstarter campaign in 2013, to raise money for my memoirs, *Borg Like Me*, and I also took great pains to keep my backers in the loop and entertained via regular campaign updates. Many backers told me it was their favorite part of the campaign and they looked forward to seeing my updates.

BE CAREFUL WITH YOUR REWARDS

Shing Yin, who's successfully launched four projects, cautions makers not to get carried away "offering up secondary products and lots of weird stretch goals. A successful crowdfunding campaign is about delivering exactly what you said you would, not delivering everyone a bunch of extra buttons and T-shirts."

Sam Brown (*Make:* contributor, game designer, and tech educator) has heard this tendency to over-promise and underestimate rewards referred to as "Kickstarter hell." He adds: "If your campaign sells 20,000 units of a reward, and it takes you a mere six minutes to box that reward, create the mailing label, apply it, weigh the package, and print and apply postage — if you multiple that by 20,000

rewards, that's a full year's worth of work for one person." Aka Kickstarter hell.

REALISTICALLY CALCULATE (THEN RECALCULATE) REWARD PRICING

Don't go crazy on too many rewards. John Dimatos, former senior director of design and technology communities for Kickstarter, says that 5–7 backer reward levels are the most common in successful campaigns, with the most typical pledge level being the $25 range, with a $100 tier level frequently delivering the most funds to a campaign.

Also make sure to nail the true price points of each. Be very thorough when calculating actual costs, including packaging, postage, and anything else that goes into fulfilling the rewards. And labor!

> The most cost-effective rewards are virtual: ebooks, apps, and other digital products that are cheap to fulfill.

DEVELOP WITH YOUR BACKERS

Many of the most successful crowdfunding creators, especially those that regularly use crowdfunding as an integral part of their business model, let their backers in on the process of developing the final product (or in defining and developing stretch goal rewards). Recognizing this, many crowdfunding sites are constantly adding tools for better creator/backer communication, offering emailed updates, project page discussion, and live video streaming and chat on project pages. The more you engage your backers in the actual creative development process, the more invested they'll be in the final product.

HAVE A THOROUGH MARKETING PLAN

This is probably one of the most overlooked aspects of crowdfunding. You put so much time into launching and maintaining the campaign, it's easy to forget that you have

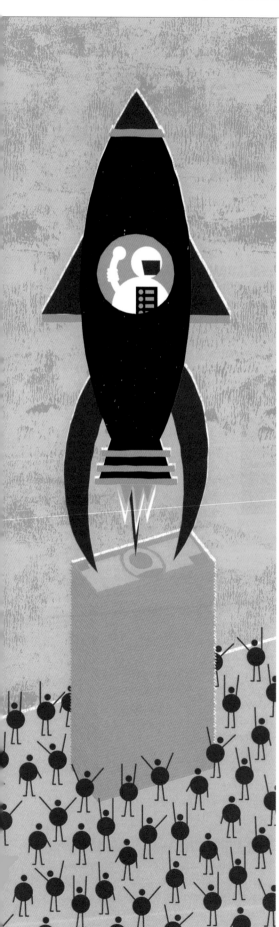

to give equal time to getting the word out, before launch and over the course of the entire campaign. Many creators plan an upfront media effort, and maybe one toward the end, but they fail to keep heat on it throughout the funding period. I thought I had a decent plan for my book project, but I could have used three times as much exposure.

BUILD BUZZ AND OFFER INCENTIVES

Many of the more successful campaigns, campaigns that I've gotten most excited about, are ones that began to build a buzz long before the actual campaign launch. For several of these, I eagerly anticipated the launch date and was excited to grab an "early bird" discounted pledge level the moment the campaign went live. Many creators do a soft launch with early bird pledge levels. They will announce the launch only to their customer base and "friends and family network," and then do most of the big PR push 24 hours (or so) later. This way, there's a run on early bird offerings from an excited base. Then, when everyone else shows up, the campaign has already put up some decent numbers on the scoreboard.

CONTEMPLATE MAKER-OPTIMIZATION

We hear plenty about Kickstarter, Indiegogo, and GoFundMe, but there are other crowdfunding sites and online communities with crowdfunding components optimized for makers and similar market small-fries. What too many creators in the maker tech domain, especially first-timers, fail to understand is the significant complexity of bringing a high-tech product, even a modest one, to market. To address this, sites have emerged that, besides helping to raise money for projects, assist you in getting your projects into production and building your market.

Probably the most successful and fully realized of these sites is Crowd Supply. Where Kickstarter doesn't want you to think of them as a store, Crowd Supply might be offended if you don't. Started in 2013, this Portland, Oregon company has served some 150 makers to date, creators like Star Simpson and her Forrest M. Mims III Circuit Classics boards, OnChip's Open-V open silicon microcontroller, and Bunnie Huang's *Essential Guide to Electronics in Shenzhen*. Crowd Supply carefully curates the projects, looking for products that "add something new and exciting to the world." Once a project is chosen, they endeavor to assist from planning and funding through development, manufacturing, and sales. They've also drafted a *Proclamation of User Rights* that they require projects to adhere to.

Baqqer is not yet as well-known as Crowd Supply. Like a lot of domain-specific services, they aim to "offer a singular experience for makers," as Baqqer founder (and former *Make:* staffer) Dan Gailey puts it. They seek to build a true maker community around the process of developing, funding, and supporting interesting projects. "One of the most important things you can do is to build a community around the things you love," says Gailey. Baqqer is still very small, and it remains to be seen how many such niche communities are sustainable, but it's heartening to see services emerging specifically targeted at makers, tech culture, and small-run high-tech products.

CONSIDER CERTIFICATION PROGRAMS

To address whether creators and projects are of sound mind and sound design, certification programs have popped up. This is a third party entity with considerable experience in tech develop and manufacturing. They examine a pending product and its funding campaign, the manufacturing timeline, and put their seal of approval on it if they like what they see. Two such programs for makers are Dragon Certified and Arrow Certified. Dragon is a fee-based program; Arrow Certified (in partnership with Indiegogo) is free to creators whose projects are approved. Arrow also sets aside "flash funding" (currently $1,000,000) to pledge in Arrow Certified campaigns.

> You may find a better solution for your goal with an alternative to crowdfunding, like Patreon, Selfstarter, and Ethereum.

TIPS FOR PROJECT BACKERS

We talked to makers who've backed numerous projects and asked for some tips that may not be as widely shared.

HONESTLY ASSESS RISK TOLERANCE

There is always some degree of risk in crowdfunding, so don't forget about the speculative nature of what you're doing. It's a lot more fun if you have realistic expectations. Don't pledge a lot of money if the project doesn't seem sound and if you're not willing to risk losing the money that you pledge.

"It's still the Wild West out there," says *Make:* contributor and prolific crowdfunding backer Kent Barnes. "Buyer beware. Kickstarter, IGG, and other sites need to pick up their game and figure out how to make crowdfunding safer for backers. If they don't, more fraud and scams are going to creep in over time." Doc Popular adds: "I'd like to see accountability become the next big trend in crowdfunding, maybe even by only offering creators part of their funding until their project is completed. That may be a little harsh, but as more people feel burned by one bad experience, it's going to fall upon the crowdfunding community to find ways of rebuilding that trust and reliability."

TAKE A CRITICAL LOOK

Ben Joffe from HAX, the well-known hardware accelerator: "Read between the lines of project descriptions and videos. You'll be surprised how much you can figure out. It's not uncommon to see that, despite all of the details, there is no proof that the creators actually created anything themselves, generally a bad sign. Sometimes, even the feasibility of the project is clearly questionable." Do your homework. Use your gut. As Joshua Lifton of Crowd Supply says: "If it seems too good to be true, it probably is."

PLEDGE SMALL, SUPPORT MANY

Kent Barnes engages in crowdfunding in a way similar to many people I spoke with. He's backed nearly 600 projects, but for many, he only pledged the minimal amount. "By pledging at the entry level, usually $1, I get the project updates. This way, you can follow the campaign and decide to raise your pledge later if you like where the project's going. I like the idea of helping, micro-financing, cool maker projects. It's a wonderful way of bringing new ideas to market and giving the money directly to a person with a bright idea." Of Kent's backed projects, 50 were unsuccessful (didn't reach their project goal) and 16 were cancelled or suspended.

Regular *Make:* contributor John Edgar Park has backed 10 projects online to date, all eventually successful except for one. "I think it's wonderful that people, particularly creative people, have the opportunity to offer highly personal, unconventional, niche projects," says Park. "Crowdfunding offers them some assurance that they've correctly gauged their audience, and it feels like a way to serve the long tail with relatively low risk. Crowdfunding allows people to create quirky and charming games, odd gadgets, and the like, without losing their shirts."

BACK TO ACCESS A PLEDGE MANAGER

Many projects these days employ a pledge manager, third-party software designed to handle the often-complex rewards fulfillment. After a successful campaign, when rewards are ready to be shipped, backers are invited into a pledge manager, and at that time, can upgrade to a higher level to receive higher rewards. By backing a project at a low level (sometimes even as low as the $1 "tip"), you can often gain access to the pledge manager. This way, you can see the entire arc of the campaign, see how many stretch goals were unlocked, and see if you still have confidence in the campaign before you invest in a significant backer level. Some creators even spell out at what level you have to pledge to access the Manager. If you're unsure, ask the project's creator.

GET INVOLVED WITH THE PROJECT

All of the major crowdfunding platforms allow you to engage with project creators. If it's a project you're excited about, get in there and help shape the final product by joining the community of backers that grows up around most projects. Project creators love this kind of input, it will help them to make a better product, and you will feel a real sense of investment in that product.

LEARN HOW THINGS ARE MADE

One interesting side benefit to crowdfunding is that it can be used as a learning process for those interested in small-scale manufacturing and how things are made. Doc Popular writes: "I think crowdfunding gives consumers a lot more insight into the products they buy. One of my favorite Kickstarter projects was for kevlar socks. The creator had done his homework, gotten quotes and leads on manufacturers, but when he went into production, he ran into one problem after another. He was good at sending out updates and now I know more about how socks are made than I ever thought possible (and the difficulties of working with kevlar thread). I felt terrible for the creator, but was glad I backed the project. Now I know more about the garments in my life and what goes into making them."

JUDGING A PROJECT'S VIABILITY

Small manufacturing facilitator Dragon Innovation created their Dragon Certified program to help provide more confidence for backers of tech projects. The idea was to create a seal of approval process from a respected source so that project backers at least know that some entity within the industry has scrutinized a potential project's mechanical, production, and market viability.

Also to help assess the viability of hardware projects being launched through crowdfunding sites, crowdfunding legal reformer (and former *Make:* Executive Editor), Paul Spinrad, has proposed the idea of a volunteer project vetting board. This group would look at proposed crowdfunding candidates and fill out a survey of pass, fail, maybe, and notes of their assessment.

Kickstarter is fond of reminding people that they are not a store. Crowdfunding is about product development and developing a market, a user-community, not just customers pre-ordering your products. As Ben Joffe of HAX, told *Make:*, "For us, Kickstarter is an 'awareness enabler.' If a campaign goes well, it can not only finance you, but also attract your investors, distributors, staff, media, and other good things." Ben continues: "We see Kickstarter as a tactic, not a company strategy (i.e. long-term success it not tied to crowdfunding results). We had companies do modestly well on Kickstarter then go on to become highly successful afterwards (e.g. Makeblock and Next Thing Co.). Inversely, success on Kickstarter doesn't mean everything will go well and doesn't 'prove' a market beyond innovators and early adopters. Real success is when a startup not only ships what it promised, but goes on to establish scalable sales and distribution." ◉

GARETH BRANWYN is a contributor to *Make:*, Boing Boing, and Wink Books. His latest book is a best-of writing collection and "lazy person's memoir" called *Borg Like Me (& Other Tales of Art, Eros, and Embedded Systems).*

Making That Matters

Nonprofit Field Ready provides disaster and humanitarian solutions in remote and low-resource areas   Written by DC Denison

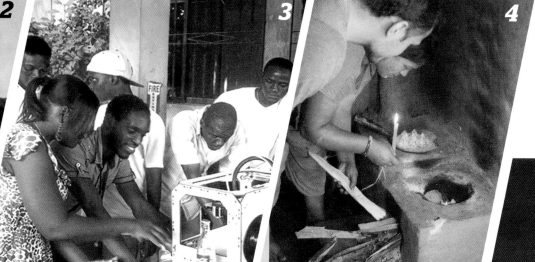

1. Mark Mellors 3D prints a water supply part at Barhabise IDP camp in Sindhupolchowk, Nepal.

2. Dara Dotz of Field Ready

3. Terry Jhonson teaches locals how to use a 3D printer in Haiti.

4. Ram Chandra tests a new stovetop burner design.

Dara Dotz is a co-founder of Field Ready, a nonprofit organization that applies maker skills in disaster areas and communities of need. This emerging practice is sometimes called "humanitarian making." Dotz, and Field Ready, have worked in hurricane-ravaged Haiti and earthquake-devastated Nepal. Dara was also a part of Made in Space, the first team to build a gravity-independent 3D printer and launch it to the International Space Station.

Q. What's your advice to makers looking to apply their skills to communities in crisis?
A. Not everyone's made to go into a disaster area. Someone might be an incredibly talented engineer, however, and we need that help. At Field Ready we work with a great community (humanitarianmakers. org) and we can plug people in, in a safe and functional way. Skype allows us to ask for help: Someone not on-site can prototype something, I'll print it in Nepal, and then give them real-time feedback. That's only going to get easier and better. Another strategy is to start in your community. Find people there who need help.

Q. How can makers get a sense of what the problems are?
A. The community will tell you. Humanitarian making is about figuring out the appropriate solution — identifying needs on the ground, working side by side with the affected population, ensuring that local people have a say in your designs and solutions, and they can use what you're making. It's important to create that sense of agency.

Q. What are some examples of humanitarian maker projects?
A. My co-founder, Eric James, worked on water sanitation and developed a new way to purify water. There's a group from Burning Man, "Burners without Borders,"

who basically build cities out of nothing. Tikkun Olam Makers runs an accessibility project to make assistive technology: they spend 3 months getting to know a person and then design a device for them.

Q. What are the important technologies in your tool kit?
A. We probably use the 3D printer most. When an earthquake hits, say, in Nepal, medical machines may break. They're antiquated — though not obsolete — and it's tough to get parts. Instead of having the clinic down for six months waiting for the part, we can use the printer to fix it. 3D printers are great for one-offs. They can sometimes solve supply chain problems.

Q. What's in your disaster maker kit?
A. My kit fits in a large suitcase and includes a 3D scanner, a 3D printer, an injection molder, a back-up computer, 3 sketch pads, and tons of pens. Having people draw with you is really helpful. I draw something: "Is this what you mean?" And they can draw a response: "No, no, it's more like this."

Q. Are there any technologies that really aren't there yet, in terms of usefulness?
A. 3D printers! They can be so powerful when they work, but in some ways they're just not ready for the field. My friends at ONO made a new 3D printer that can fit in a purse and work off a cell phone — for $99. That could be completely revolutionary!

Q. Anything on the horizon that you think will make a difference?
A. The real power isn't with us, but how it works when we give it to others. I'm always looking for new technology and figuring out how to cross it over to the rest of the world. I call it Moore's law table scraps. Once the price of a tool comes down, I like to test it, see if it works for people, and then hand

it to them. We're developing search and rescue tools, like a concrete lifter to pull victims out of collapsed buildings for a fraction of the cost of the traditional solution. It's built from materials found on-site, like reclaimed car mats. So far we've been able to lift 13 tons!

Q. Laser cutters get a lot of action in makerspaces. Are they valuable in the field?
A. When you're thinking about different tools, you really want to pay attention to what people need. Sometimes amazing tools aren't suited for field work. Someone might say, "CNC machines are so cool, we can use them to build houses in Haiti!" But there are no trees in Haiti, so you can't get wood. Do you really want to bring a CNC machine there? Millions of people are out of work in Haiti. For less than the cost of a CNC machine, you could hire 100 people and give them work for 3 months.

Q. In addition to working in disaster areas, you've worked on Made in Space, putting manufacturing projects into orbit. Do they have anything in common?
A. One common thread is the supply chain challenge. You can imagine how expensive and dangerous it is to deliver things to space. You put all your supplies on a giant explosive device that costs millions of dollars to get to this one point in orbit, and you hope it docks without blowing up. Going into a disaster zone can be similar — people are in a dangerous place that's expensive to get supplies to, you get robbed, the environment might be polluted. In both situations, we're trying to figure out how to build resiliency in the extremes. When humans are cut off from the rest of humanity, how do we support them, and what are the tools that they need to have agency and push through? ◐

Three organizations you can connect with today:
humanitarianmakers.org
tomglobal.org
fieldready.org

DC DENISON is the co-editor of the *Maker Pro Newsletter*, which covers the intersection of makers and business, and is the senior editor, technology at Acquia.

Read the full interview, and find more Maker Pro news and interviews at makezine.com/go/makerpro.

Written by Tim Deagan

WORKSPACE FIRE SAFETY

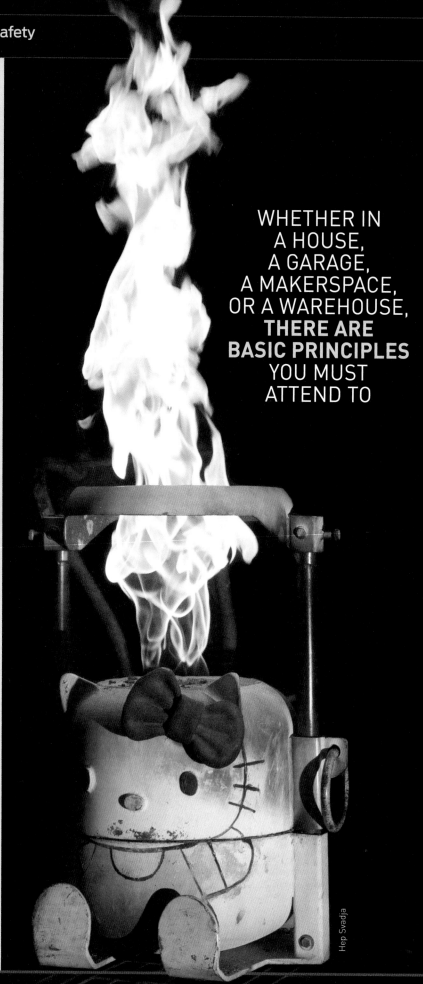

WHETHER IN A HOUSE, A GARAGE, A MAKERSPACE, OR A WAREHOUSE, **THERE ARE BASIC PRINCIPLES** YOU MUST ATTEND TO

TIM DEAGAN

(@TimDeagan) casts, prints, screens, welds, brazes, bends, screws, glues, nails, and dreams in his Austin, Texas shop. A career troubleshooter, he designs, writes, and debugs code to pay the bills. He's the author of *Make: Fire*, and has written for *Make:*, *Nuts & Volts*, Lotus Notes Advisor, and Database Advisor.

Hep Svadja

THE RECENT TRAGEDY AT A DIY VENUE IN OAKLAND KNOWN AS GHOST SHIP IS A heartbreaking call to action on the part of all makers. Thirty-six people lost their lives when a fire broke out during a music performance and the only exit, a staircase constructed from pallet wood, became impassable. Fire is one of humanity's oldest tools, but it still requires vigilance and preparation to keep it from becoming a terrifying and destructive force. As the Flame and Safety Coordinator for Maker Faire Austin, I spend a lot of time thinking about how to encourage creativity while keeping everyone safe.

As we extend the bounds of who is a maker and where making occurs, we must also expand our safety practices to protect ourselves and others. Build co-ops, makerspaces, garages, and Maker Faires have all increased over the last few years. Many events occur at DIY venues that draw crowds for education or entertainment. We're entrusted with making sure that the people visiting or working at our spaces are safe.

Whether you work in a house, a garage, a makerspace, or a warehouse, there are basic principles of fire safety that you must attend to: ignition sources, fuels, response, and egress. Each of these is a topic that can be discussed in book length, but we'll talk about their fundamentals so that you can review how they apply to your workspace.

> There are a lot of people thinking and writing about this; I can't recommend highly enough Gui Cavalcanti's piece, "A Guide to Fire Safety in Industrial Spaces" (makezine.com/go/industrial-space-fire-safety).

IGNITION SOURCES

Ignition sources can be obvious, like open flames or heating elements. However, subtle or invisible hazards — like friction, chemical reactions, electrical resistance, and thermal radiation — can be just as deadly. Fires can start from a variety of unexpected sources, such as:

» Power tool blades that get hot from the friction of cutting
» Fiberglass resin with excessive catalyst that heats up too quickly
» Items that concentrate sunlight through reflection or refraction
» Devices that overheat wires or cords

Ignition sources can also come from old or worn building infrastructure. According to the National Fire Protection Association (NFPA), the leading cause of accidental fires in warehouses is arcing from electrical distribution and lighting equipment.

While it's impossible to list every potential ignition source, most can be addressed by considering your operations, storage, and standards. What are possible sources of heat (Figure )? Are flammable materials and "works in progress" stored appropriately? Are you overloading a wall socket? (And are the devices and the circuits powering them appropriate for the ampacity and duty cycle?) Look around your workspace and start asking yourself these kinds of questions.

FUEL

When an ignition source meets a fuel source, fire results. Flammable items are everywhere: fuels, fabrics, solvents, cleaning and painting products, wood, paper, plastics ... the list is huge. Eliminating all possible fuels isn't feasible, but managing their exposure to ignition sources is (Figure ). Examine work practices with an eye toward how accessible fuels (and worse, fuel vapors) are to ignition sources.

It's easy, especially in crowded workspaces, to have a jumble of tools and fuels. When working in a hurry, it's easy to lose awareness of where you tossed that acetone rag. The hot cinder from your cutting wheel might smolder for hours before erupting into flame in a pile of sawdust. In a cold workspace it's hard not to want to seal up the windows and doors to keep in heat while allowing flammable vapors to concentrate. But maintaining a conscious segregation of heat sources and fuels is not optional.

Take a walk through your space and try to spot anything that could burn. Then exercise your imagination and picture how it could get ignited. Some things will be highly unlikely, but others will surprise you. Organize your space to reduce risk (Figure ).

FIRE RESPONSE

Having a means to detect a fire, typically by its smoke, and raise an alarm is an absolute requirement. No workspace of any size should be without smoke alarms. Resist the urge to turn detectors off during an

Scrap wood is handy for projects, but is tasty fuel for a fire. Make sure it's stored far away from possible ignition sources.

Anticipate how you will store and dispose of flammable items, like oily rags, before you begin working on a project.

CHOOSING EXTINGUISHERS

In the United States, fires are categorized under five types:

A: Ordinary solid combustibles (A for "Ash")
B: Flammable liquids and gases (B for "Barrel")
C: Energized electrical equipment (C for "Current")
D: Combustible metals (D for "Dynamite")
K: Oils and fats (K for "Kitchen")

There are many types of extinguishers available, with different capabilities. Here are the most common:

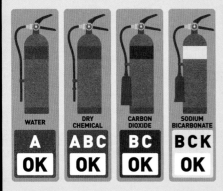

WATER	DRY CHEMICAL	CARBON DIOXIDE	SODIUM BICARBONATE
A	ABC	BC	BCK
OK	OK	OK	OK

» **Pressurized water extinguishers.** Great for type A fires, but extremely dangerous when used on electrical or grease fires.

» **ABC Dry Chemical.** This is the most common extinguisher. It's rated for three types of fire, but uses a corrosive agent, monoammonium phosphate, that has some health risks.

» **CO_2 extinguishers.** Excellent for type B and C fires, but can have trouble putting out Type A fires. Plus, they're expensive.

» **Sodium Bicarbonate extinguishers.** Good alternatives for B and C fires and are starting to be marketed for Type K kitchen fires. They are less hazardous than monoammonium phosphate, so they're not as offensive to spray near people, but they aren't rated for Type A.

No one type of extinguisher is perfect, but it's better to over-respond than under-respond. If you have time to assess the type of fire and can safely use the extinguisher specific to the fire type, that's a great choice. However, a dry chem ABC is the safest bet if you're unsure.

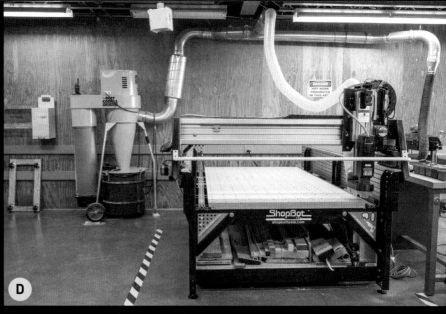

When you install a heavy-duty machine in your space, make sure you also account for proper ventilation.

A fire extinguisher should be easily seen from eye level and readily accessible.

operation that might cause a false alarm. It is better to increase ventilation (Figure **D**) or move the offending operation outdoors.

Facing a fire is no time to start wondering how to respond to one. Outfit your workspace with appropriate fire extinguishers in easily accessible and highly visible locations (Figure **E**). Be sure that your extinguishers are ready and that you're using them correctly before a fire occurs. Hold safety sessions with the makers in your space so they know what to do in a fire or other emergency.

EGRESS, AKA HOW TO GET EVERYONE OUT

Smoke, flames, and fear create an environment that's difficult to navigate. Visualize how people will escape your workspace if accidents occur in different areas. The worst-case scenario is when the only exit is blocked. If you can't provide egress through another door or window, you must organize your space to leave your single exit clear. It is also crucial that exits be obvious even in low visibility or during power outages to help people who may be unfamiliar with your space get out safely in an emergency.

A corollary to egress is access for emergency personnel. How hard will it be for responders to get to the space with hoses and other emergency equipment? Does parking get so crowded that fire trucks or ambulances can't get close

Adobe Stock, Hep Svadja

Ensure exit signs will illuminate even if the building loses power during an emergency.

enough to help? You may never need emergency services, but hoping that will be the case is no substitute for preparing to make their job easier if you do.

MATCHING SAFETY TO SPACE

The basic concerns about fire safety are universal, but different spaces can have different needs. The core threats are similar, but the characteristics and priorities of concern can change.

Small adjunct spaces such as garages and residential workshops typically add additional risk due to housing many functions, tools, and supplies into a cramped space. This increases the importance of keeping ignition sources separate from fuels. Various types of hot work may need to be moved outside.

Medium-sized spaces like offices and prototyping shops in industrial parks can have a mix of top concerns. In offices, the primary focus is getting everyone out of the building safely and efficiently. If you work in an office, know where you and your office mates should meet if evacuated so that you can be sure everyone got out OK. In small industrial spaces like cabinet shops, CNC fabs, and other places where tools or hot work is going on, evacuation plans are essential, but easy access to extinguishers so that small fires can be stopped in their tracks starts to become more of a priority.

Large warehouse spaces often have special requirements. As noted, aged or damaged electrical wiring is the leading cause of accidental fires in warehouses. Large spaces also often have nooks and crannies where old equipment, or, worse, chemicals and paints are stored. Along with egress and extinguishers, regular inspections of the space to make sure that structural problems aren't brewing starts to become important. ⊘

BALANCING
Artistic Freedom
AND NECESSARY REGULATION

One of the most painful aspects of the Oakland Ghost Ship fire is that it's easy to list all the things that might have kept the fire from occurring, or at least from being as deadly. The painful truth is that the risks were dramatically increased by the uses the space was put to, which were created outside of codes, regulations, and permits. As I write this, as a Maker Faire Safety Coordinator, Safety Officer at my job, owner of an 80-year-old church we converted into a home, author of a book on building safe fire art, and former volunteer firefighter, I feel like a hypocrite.

In my 20s I lived in a warehouse (pictured above) that was almost exactly like the Ghost Ship space — one of three arts warehouses that I lived and worked in through the years. I believed then, and I believe now, that the output from these spaces is a phenomenal creative engine for a vibrant society. The young me does not want the old me to forget the value of the creative culture operating under the radar.

There are thousands of these kinds of spaces across North America. Could the residents or artists in these spaces get contractors or inspectors to help them understand the space's risks? Sure, but it would generally expose them to the legal requirements of working within code. And that brings price tags that range from tens to hundreds of thousands of dollars.

The people who can afford to do that are often considered gentrifiers. They are frequently seen as driving out creative but low-income artists and makers.

Does this mean that we should turn a blind eye to situations where people are living or working unsafely? No, it can't. We have to take care of each other even when it means limiting possibilities to limit risk. Does it mean we should give up on places that serve artists and makers who don't have money? Deep in my bones I believe it doesn't have to.

There are examples of a middle ground. There are co-ops and collectives that pool resources to create large open workspaces. There are membership-based makerspaces and places like the Burning Flipside Warehouse in Austin where the burn community shares resources. They may have to conform to codes, which limits some activities, and they might not be residential, but they are open and willing to share space and resources.

If you want to respond to the tragedy that befell Oakland, think about how you can help make more spaces for everyone to create in, safely. It's important that we create access along with art, that we share opportunity as well as knowledge. As a community we have to adopt safety as a way to care for one another, not just a necessary burden.

Hep Svadja, Tim Deagan

Written by Jose Gomez-Marquez • Illustrated by Colin Johnson

BIOZONE

The DIY life sciences arena is a broad, innovative world that's perfect for makers

EVER SINCE THE SUMERIANS LEARNED TO HACK YEAST TO MAKE BEER, we've enjoyed the growth of the biological construction set. But beer doesn't make a civilization alone. Fermentation experiments led to biochemistry that describes biomolecules. Thanks to microscopes we learned germ theory and how disease is transmitted.

This biological construction set is now by far the biggest of all forms of making. Typical electronics makers have about 150 types of components. In contrast, chemists have access to over 20 million synthetic chemicals, with 1 million new ones each year. There are more than 3 million antibodies on the CiteAb Search engine. That's just a couple of categories. Then there's health making: fabricating hardware that aids the human body and our health.

We can call this broad grouping "Life Sciences Making" — a big-tent term for DIY biology, maker health technologies, maker bionics, DIY molecular gastronomy, and other fields where your bits and atoms also include cells and life. It's the new kid on the block with a very old legacy.

MEDICAL MAKING

Medical technology was shaped by makers. Dr. John Heysham Gibbon created the heart-lung machine in 1953 and open sourced the blueprints for others to DIY. Margaret Crane, a graphic designer, hacked together a home pregnancy test in 1967. Our team at MIT discovered a community of underground maker nurses that have been hacking medical solutions since 1905. They even had their own *Make:*-like publication that shared how-tos, blueprints, and hardware they made for patients, sustaining a vibrant community in health care.

Over time, industry slowly black-boxed our medical technologies and discouraged makers from participating — but they never went away. Today, life science and health makers form networks like MakerHealth

JOSE GOMEZ-MARQUEZ is a Honduran-born scientist who leads the Little Devices Lab at MIT and is a co-founder of MakerHealth and MakerNurse.

(makerhealth.co) and DIY bio communities. They fuel open protocols and cheap instrumentation.

Need a place to begin? Start with a teardown and remake what you see at the doctor's office. There are more than 6,000 Nerf blaster project entries on Instructables. That device ($15) is more complex than its medical cousin, the EpiPen self-injector ($650) — put the two ideas together and you have a disruptive DIY project ready to go. A bike-powered nebulizer ($10) works just as well as the store-bought one ($80). The Nightscout and the OpenAPS communities saw continuous glucose monitors and insulin pumps the same way OpenSprinkler did sprinkler controllers and created an open source artificial pancreas — you can find the instructions on Facebook! The Open Source Malaria project demystifies drug development for global health so that even high schoolers can synthesize life-saving medication. OpenBCI brings the dark arts of physiological signal monitoring and control to the masses. It goes on and on.

Not all teardown are easy. Antibodies all look the same in a vial: clear, colorless, and boring. It's just how nature makes them. Imagine doing an electronics teardown only to find that all of the components lack labels and look the same. You need instrumentation. Enter groups like DIYbio, Prakash Lab, Tekla Labs, and science journals like PLoS and HardwareX that work on tools like affordable PCR machines, centrifuges, $1 microscopes, laser-cut gel electrophoresis boxes, and Arduino spectrophotometers to name a few.

COLLABORATION IS KEY

Recall that the reason we have maker drones, affordable microcontrollers, and even democratized cooking is that after a few people created tools, they created communities, shared plans, and collaborated in the open. I often tell the

nurses we work with that they'll be more successful if they create and share a constant stream of things, instead of holding on to a one-hit wonder. This is how makers are going to overcome black-box medical technology. And it's exactly what the Open Insulin Project at Counter Culture Labs does (page 35). It's making its biological building blocks open. Insulin is an old discovery but sold at runaway prices. The OIP is working on an open version of insulin that may one day reach the 400 million diabetics globally.

Successful belief systems have many cathedrals. I gave up on trying to categorize every type of biohacking, medical making, and cellular tinkering a while ago. More communities mean more parts in the construction set. At the Little Devices Lab here in MIT, we make hackable construction sets so patients, doctors, and nurses can make their own medical devices: Smart asthma inhalers, LED-powered diagnostic gels, and a Lego-like platform for creating biochemical reactions called Ampli (see Ampli: Reinventing the Biochemistry Set). We create makerspaces in hospitals so that a nurse and a doctor are just an elevator ride away from 3D printers and more specialized tools like IV makers and patient monitor development systems. A typical day includes medical teardowns, prescribing prototypes made with patients, and exploring ways to make the next "medical Arduino."

A HEALTHY FUTURE

Will we download an .stl file to print a pill? Or make Arduino-powered glucometers with DIY test strips? Or flush a glowing biocircuit into the water table? These scenarios are not only possible today — they are feasible with less than $100 in materials. Policy is catching up to this democratization of tools. As makers, we have a duty to create medical and biological parts that demonstrate a public benefit.

Makers gave us the modern day techniques of medicine and biology. Nature gave us the parts. While protocols got complicated, our ability to tinker, to question, to democratize experimentation never went away. Today's life sciences tools and communities will reinvent the way cells say "Hello World" and offer patients a prescription of a prototype that can save their life. ◗

AMPLI: REINVENTING THE BIOCHEMISTRY SET

We at Little Devices Lab wanted to democratize the way modern day assays involving antibodies, reagents, and living parts work in complex reactions to make things like diagnostics, cell cultures, and molecular biology circuits. Ampli works by taking what normally is done in a lab flask with pipettes, and transforms these protocols into literal building blocks powered by solid-state flow. It's like a breadboard, but for biology. The interlocking Lego-like blocks form chemical and biological circuits connected by a liquid flow. So far, users have made Zika and cancer diagnostics, nanoparticles, and even perfume! ampliscience.com

Nikolas Albarran, American Journal of Nursing, Hep Svadja

Written by diybio.org

BIOHACKING SPACES Near You

Want to get hands-on with science? DIYbio.org has assembled this extensive list of bio labs and meetup groups around the world. Look closely, there may be a lab or meetup in your area where you can get started on your project.

USA-EAST

Asheville DIY Bio Meetup
Asheville, NC
meetup.com/
Asheville-DIYBio

Baltimore Under Ground Science (BUGSS)
Baltimore, MD
bugssonline.org

Capital Area BioSpace (CABS)
Bethesda, MD
meetup.com/
CapitalAreaBioSpace

Boston Open Science Lab (BosLab)
Boston, MA
boslab.org

Genspace
Brooklyn, NY
genspace.org

MIT DIYbio
Cambridge, MA
openwetware.org/wiki/
MIT_DIYbio

Open Bio Labs
Charlottesville, VA
openbiolabs.org

DIYbio South Carolina
Columbia, SC
facebook.com/diybiosc

Cap City Biohackers
Columbus, OH
capcitybiohackers.org

Ronin Genetics
Durham, NC
roningenetics.org

Great Lakes Biotech Academy
Indianapolis, IN
greatlakesbiotech.org

Try Sci
Kansas City, MO
trysci.org

DIYbio Madison
Madison, WI
meetup.com/diyBio-Madison

MN DIYbio
Minneapolis, MN
meetup.com/MN-diyBio

Harlem Biospace
New York City, NY
harlembiospace.com

Biologik Labs
Norfolk, VA
biologiklabs.org

FamiLAB
Orlando, FL
familab.org

Triangle DIY Biology
Research Triangle Park, NC
tridiybio.org

USA-WEST

Berkeley BioLabs
Berkeley, CA
berkeleybiolabs.com

Bio, Tech and Beyond
Carlsbad, CA
biotechnbeyond.com

Denver Biolabs
Denver, CO
denverbiolabs.com

La Jolla Library Bio Lab
La Jolla, CA
lajollalibrary.org/your-library/
bio-lab

Biodidact
Los Alamos, NM
biodidact.net

TheLab
Los Angeles, CA
thel4b.com

Counter Culture Labs
Oakland, CA
counterculturelabs.org

PortLab
Portland, OR
portlabdiy.org

DIYbio San Diego
San Diego, CA
meetup.com/
DIYbio-San-Diego

Wet Lab
San Diego, CA
wetlab.org

Indie Bio
San Francisco, CA
sf.indiebio.co

HiveBio Community Lab
Seattle, WA
hivebio.org

BioCurious
Sunnyvale, CA
biocurious.org

CANADA

Brico.Bio
Montreal, QC
brico.bio

Nelson-BC-DiyBio
Nelson, BC
nelson-bc-diybio.weebly.com

BioTown
Ottawa, ON
biotown.ca

DIYbio Toronto
Toronto, ON
meetup.com/DIYbio-Toronto

Open Science Network
Vancouver, BC
opensciencenet.org

EUROPE

ABiohacking
Albacete, Spain
facebook.com/groups/
ABiohacking

Waag Society's Open Wetlab
Amsterdam, Netherlands
meetup.com/Dutch-DIY-Bio

DIY Bio Barcelona
Barcelona, Spain
diybcn.org

Biotinkering Berlin
Berlin, Germany
biotinkering-berlin.de

Open BioLab
Brussels, Belgium
openbiolab.be

Bio.Display
Budapest, Hungary
biodisplay.tyrell.hu

Biomakespace
Cambridge, UK
biomake.space

Biologigaragen
Copenhagen, Denmark
biologigaragen.org

DIYbio Ireland
Cork, Ireland
groups.google.com/
forum/#!forum/diybio-ireland

Bio Art Laboratories
Eindhoven, Netherlands
bioartlab.com

Bioscope
Geneva, Switzerland
bioscope.ch

ReaGent
Ghent, Belgium
reagentlab.org

Open BioLab
Graz, Austria
facebook.com/
OpenBioLabGraz

DIYbio Groningen
Groningen, Netherlands
diybiogroningen.org

Biotop Heidelberg
Heidelberg, Germany
biotop-heidelberg.de

DIYbio Kiev
Kiev, Ukraine
groups.google.com/
forum/#!forum/diybio-kiev

L'Eprouvette
Lausanne, Switzerland
eprouvette.ch

Hackuarium
Lausanne/Renens,
Switzerland
wiki.hackuarium.ch

London Biohackspace
London, UK
biohackspace.org

London Hackspace
London, UK
london.hackspace.org.uk

BioChanges
London, UK
meetup.com/BioChanges

Symbiolab
Maribor, Slovenia
irnas.eu/symbiolab.html

Biohacking
Moscow, Russia
vk.com/biohax

Biogarage
Munich, Germany
biogarage.de

DIYbio Belgium
Namur, Belgium
diybio.be

OpenGenX
Nottingham, UK
opengenx.wordpress.com

La Paillasse
Paris, France
lapaillasse.org

Project Biolab
Prague, Czech Republic
brmlab.cz/project/biolab

BioNyfiken
Stockholm, Sweden
bionyfiken.se

Hackteria
Switzerland/Slovenia
hackteria.org

Be.In.To
Turin, Italy
facebook.com/be.into.7

ASIA

F.lab
Bangkok, Thailand
facebook.com/
FLabDIYbioThailand

DIYio Hong Kong
Hong Kong, China
meetup.com/DIYBIOHK

BioRiiDL
Mumbai, India
bioriidl.org

DIYbio Singapore
Singapore
diybiosingapore.wordpress.
com

Hackuarium

DIYbio Israel
Tel-Aviv, Israel
groups.google.com/
forum/?fromgroups#!forum/
diybio-israel

BioHubIL
Tel-Aviv, Israel
facebook.com/
groups/1725450001000736

BioClub
Tokyo, Japan
bioclub.org

LATIN AMERICA

DIYbio Mexico
Guanajuato, Mexico
facebook.com/groups/
DIYbioMexico

Biomakers Lab
Lima, Peru
facebook.com/
groups/547202812114071

SyntechBio Network
Sao Paulo, Brazil
syntechbio.com

Garoa Hacker Club
Sao Paulo, Brazil
garoa.net.br/wiki/Biohacking

Synbio Brasil
Sao Paulo, Brazil
synbiobrasil.org

OCEANIA

BioHackMelb
Melbourne, Australia
facebook.com/
groups/698017880316967

DIYbio Perth
Perth, Australia
facebook.com/groups/
diybioperth

BioHackSyd
Sydney, Australia
meetup.com/biohackoz

DIYBIO.ORG was founded in 2008 with the mission of establishing a vibrant, productive and safe community of DIY biologists. They believe that biotechnology and greater public understanding about it has the potential to benefit everyone.

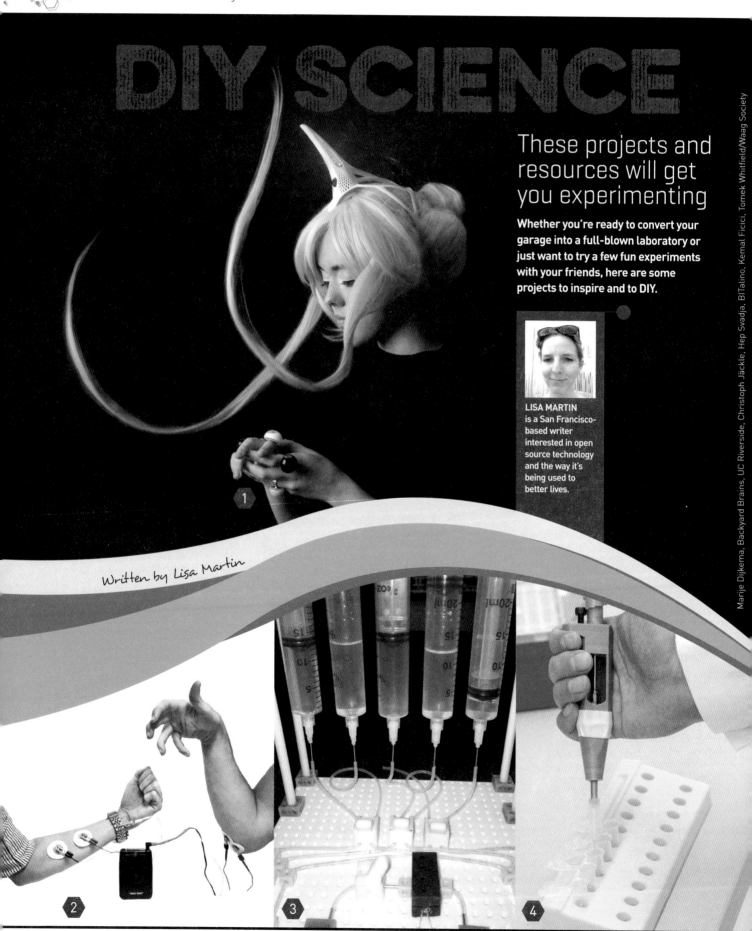

DIY SCIENCE

These projects and resources will get you experimenting

Whether you're ready to convert your garage into a full-blown laboratory or just want to try a few fun experiments with your friends, here are some projects to inspire and to DIY.

LISA MARTIN is a San Francisco-based writer interested in open source technology and the way it's being used to better lives.

Written by Lisa Martin

Marije Dijkema, Backyard Brains, UC Riverside, Christoph Jäckle, Hep Svadja, BITaino, Kemal Ficici, Tomek Whitfield/Waag Society

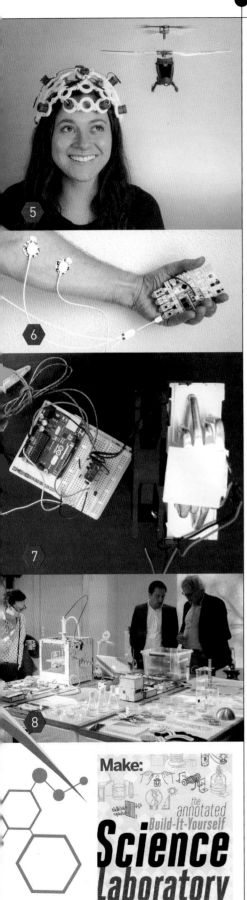

❶ AGENT UNICORN
iq.intel.com/unicorn-wearable-uses-neuroscience-to-help-kids

Anouk Wipprecht designed this playful unicorn horn to aid ADHD researchers by making EEG sensors a more comfortable and playful experience for children. The sensors track changes in brain waves that indicate increased focus, which then triggers cameras to start recording the experience.

❷ HUMAN-TO-HUMAN INTERFACE
makezine.com/2015/05/13/use-emg-control-friends

Makers are exploring a growing number of neural interfaces to control devices, but Greg Gage of Backyard Brains uses electromyography, or EMG, to control something more interesting than a robot — he's controlling humans! With human-to-human interfacing you can send signals from your muscles to your friend's and make them move under your command.

❸ LEGOS FOR BIOCHEMISTRY
makezine.com/go/biochemistry-tool-blocks

Researchers at the University of California, Riverside created a Lego-like system of building blocks that can be used to create biological or chemical instruments on the fly. The blocks, known as Multifluidic Evolutionary Components (MECs), serve simple functions like valving, pumping, mixing, controlling, and sensing, and can be rearranged to serve different functions.

❹ 3D PRINT YOUR LAB GEAR
makezine.com/go/3d-printed-lab-gear

3D-printed lab equipment is a great way to begin setting up your home lab. There is a wide variety of open source designs with the citizen scientist in mind, including Raspberry Pi microscopes, centrifuges, syringe pumps, and much, much more — this journal article lists a full range of the options available to makers.

❺ BRAINCOPTER
makezine.com/go/openbci-braincopter

OpenBCI's boards and components allow amateur scientists to easily and affordably access the EEG, EKG, and EMG (brain, heart, and muscle) activity of a subject. Through its open protocol, you can track health data and brainwave activity and use those measured values to trigger and control any number of other programs or projects. One cool implementation: controlling a toy helicopter throttle using just your brain's alpha waves.

❻ BIOHACKING BOARD
makezine.com/projects/use-bitalino-graph-biosignals-play-pong

The BITalino (r)evolution is a microcontroller with built-in biosignal sensing modules, making it perfect for transforming your body into input. Use this board to play the classic pong video game by moving your arm to move the paddle, or get creative using input from your heart activity, brain waves, skin conductance, or muscle signals.

❼ REPLICATE DNA AT HOME
makezine.com/go/replicate-dna-home

Being able to replicate DNA is a necessary part of experiments that involve mapping genomes, detecting viruses and bacteria, and diagnosing genetic disorders, but the lab equipment that enables this replication can cost upwards of $10,000. This DIY polymerase chain reaction (PCR) thermocycler setup is an affordable alternative targeted for high schoolers.

❽ OPEN SOURCE BIOHACKING CLASS
waag.org/en/project/biohack-academy-biofactory

BioHack Academy offers a 10-week online course that covers the principles of biotech and how to construct 14 pieces of DIY lab equipment. Even if you don't join one of the sessions, the group has videos from past classes up on their site and the files for the lab equipment on Github.

❾ BUILD-IT-YOURSELF LABORATORY
makershed.com/products/the-annotated-build-it-yourself-science-laboratory

If you were building your own science lab in the 1960s, you were probably using Raymond E. Barrett's wonderful resource to equip it. This updated version includes Barrett's original plans for equipment and experiments along with modern suggestions from Windell H. Oskay of Evil Mad Scientist Laboratories. ◷

Written by Lisa Martin ● Illustrated by James Burke

DIABETES AND DIY

Makers in need are accessing artificial pancreas technology to improve their lives

LISA MARTIN
is a San Francisco-based writer interested in open source technology and the way it's being used to better lives.

DANA LEWIS WANTS TO IMPROVE THE LIVES OF DIABETICS, AND HAS SPENT THE LAST THREE YEARS WORKING TO PERFECT A DIY ARTIFICIAL PANCREAS SYSTEM — and making the plans available and easy for others to implement.

The project, OpenAPS, allows a person to use the data from their continuous glucose monitor (CGM) with a small computer like a Raspberry Pi or Intel Edison to make micro adjustments to basal insulin being delivered through a pump. Connected apps allow a person to let their system know when they have done anything (like sit down for a hearty meal or run a marathon) that might make their glucose jump or fall unexpectedly. It makes managing diabetes more automated with the goal of making blood sugar levels more stable over time.

> **IT SOUNDS DRAMATIC TO SAY THAT LIVES ARE AT STAKE ... BUT THEY ARE.**

CAUSE FOR ALARM
Lewis, as a Type 1 diabetic living on her own, needed a CGM with an alarm that would be loud enough to wake her if her glucose levels dropped too low while she slept. Turning up the volume on an alarm seems like a simple request, especially considering that the consequences of it not waking her up could be a hypoglycemia-related death, but Lewis met resistance.

"I kept asking the device manufacturers for louder alarms," she says. "The manufacturers usually responded, 'the alarms are loud enough, most people wake up to them!' This was frustrating, because clearly I'm not one of those people."

> **THE GOAL IS TO IMPROVE ACCESS TO ARTIFICIAL PANCREAS TECHNOLOGY.**

Also frustrating was the fact that for the longest time she couldn't access her own medical data from her CGM in real time. If she had that, she knew it would be simple enough to make a custom alarm.

TAKING CONTROL
In November 2013, by chance, she found John Costik tweeting about how he'd managed to get the CGM data from his son's device. That tweet gave hope to

Lewis, and other patients and their loved ones, and helped spark a movement: #WeAreNotWaiting — for those unwilling to sit tight until commercial tools become available. So she reached out.

With help from Costik, and an active open source community sharing information on how to access the functionality of CGMs and insulin pumps, Lewis was able to really get started. Within a year she and her boyfriend (now husband) Scott Leibrand moved from creating customizable alarms to creating algorithms to read data from her CGM and send the correct commands to her insulin pump to make proactive dosage adjustments, closing the loop and creating her first DIY pancreas.

With OpenAPS, this went from being a personal project to being an open source, community-focused effort. "The goal that Scott and I built the OpenAPS community around is to improve access to artificial pancreas technology," Lewis says. On their site openaps.org, they've provided easy to understand documentation for setting up OpenAPS and have encouraged users to contribute, ask questions, and help expand OpenAPS compatibility to other devices.

A CALCULATED RISK
Apart from the existing diabetes devices (the CGM and insulin pump), the system is 100% DIY. "There are numerous ways to customize it, and you can use all the Lego pieces or sub in your own," Lewis says. "This means you can use a community-built and -vetted algorithm, or create your own algorithm to drive the decision making for the system." It is also worth noting that the system is not a "set and forget" setup; according to the OpenAPS website: "You'll still be actively managing your diabetes and doing basic self-care as you were before — this includes everything from meal boluses, checking BG and calibrating the CGM, changing out pump sites, etc."

Since the system is not FDA approved, people interested in using it must make it

Early OpenAPS rig with an insulin pump, continuous glucose monitor, and Carelink USB stick.

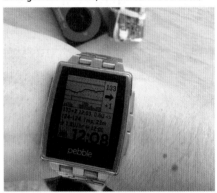
Dana Lewis' Pebble watch showing blood glucose levels and artificial pancreas activity.

Handmade maple box to hold OpenAPS artificial pancreas, closed for carrying around in a purse.

Simply slide the cover open and plug it in to charge the battery at night.

Hep Svadja

One OpenAPS user's rig: Phone to monitor trends; receiver to transmit blood sugar data; insulin pump; RileyLink to remotely control pump.

DIABETES-INSPIRED HEALTH HACKING

NIGHTSCOUT

nightscout.info

Nightscout is an open source project focused on allowing CGM users access to their blood sugar data in real time by putting that information on the cloud. Besides being behind the browser-based visualizations for OpenAPS, it can also be used to review data from a phone or smartwatch, or monitor children with Type 1 diabetes remotely.

EPIPENCIL

makezine.com/go/epipencil

EpiPens are used to administer emergency doses of epinephrine in the event of a life-threatening allergic reaction. When the price for an EpiPen suddenly increased to $300, Michael Laufer created a video showing the world how to make the EpiPencil using supplies more commonly used to treat diabetes. Not including the cost of the epinephrine (for which you would need a prescription) the EpiPencil costs less than $35.

DIY'S NOT RIGHT FOR EVERYONE, BUT HAVING THE CHOICE IS A HUGE STEP ABOVE HAVING TO WAIT YEAR AFTER YEAR AFTER YEAR.

for themselves. Using unregulated devices is a risk, but many OpenAPS users feel it's a chance worth taking. "It sounds dramatic to say that lives are at stake ... but they are," Lewis says. "People unfortunately still have a risk of dying in their sleep in the short term due to hypoglycemia, and are at increased risk long term for diabetes complications from hyperglycemia."

As I write this there are at least 177 people using some form of OpenAPS to manage their diabetes, and as a community they've logged "an estimated 550,000 hours of real-world closed loop experience."

FREEDOM OF CHOICE

"[Technology] has progressed so that we now have a choice to wait for a commercial system, or not," Lewis says. "I'm proud that because of the OpenAPS community, those with access to the compatible devices can make the choice to DIY or to wait for a commercial system. It's not right for everyone, but having the choice is a huge step above having to wait year after year after year. (For context, I've been living with Type 1 diabetes for 14+ years and have been hearing about this technology being 'a few years away' for most of that time.)"

After so many years of waiting the first (and only) commercially available and FDA-approved closed loop system for diabetics is actually due spring of 2017. This does not, however, spell the end for OpenAPS. "Even when one or two systems become commercially available, that doesn't mean they'll be perfect — or accessible to all," Lewis explains. "As a community, we still have work to do to help manufacturers advance the features of these systems and make them more quickly available, and more accessible." ●

THE OPEN INSULIN PROJECT

Biohackers are opening the door for generic medication production

Written by Lisa Martin

Hep Svadja

MEDICAL PATENTS TYPICALLY LAST 20 YEARS. Those for insulin, however, have been maintained for nearly a century due to minor yet regular advancements to its production process. The team behind the Open Insulin project now hopes to change that by creating a new production protocol and sharing it so a generic drug company can produce a low-cost version of the compound that diabetics have relied on since 1922.

"The state of the art in diabetes treatment has changed little in decades, something that I personally am frustrated by along with many others living with diabetes," explains Anthony Di Franco, one of the Open Insulin organizers. "Whether directly or indirectly, we're hoping our work will improve access to insulin."

Di Franco has had Type 1 diabetes since 2005. His initial interest in hacking diabetes was in closed-loop glucose systems and DIY pumps. The idea of making a bioreactor to create insulin seemed like a remote possibility in 2011 when he co-founded Oakland, California-based biohackerspace Counter Culture Labs. Then, in spring of 2015, Di Franco was introduced to Isaac Yonemoto, who has a background working on insulin, and Arcturus BioCloud, a biotech startup that could provide DNA synthesis services. This made Open Insulin seem like an achievable goal. They formed a meetup group, successfully crowdfunded their experiments, and began the lab work by January of 2016.

"The main method we're looking at involves expressing human proinsulin in E. coli and cutting and folding it into insulin using a series of steps involving treatments with enzymes and chemicals that parallel what's done in the body," Di Franco says.

So far the group has succeeded in creating proinsulin and are working on confirming that it was made correctly. The next steps will be cutting and folding the proinsulin to make insulin. They don't expect to initially be able to produce insulin that is pure enough or scalable enough to be manufactured, but at that point they plan on seeking a partnership with "an established small manufacturer or academic lab equipped to address these aspects properly."

"We're aiming to set a milestone in what can be done by biohackers in a modestly funded and equipped lab, and inspire others to take on more ambitious projects and share the knowledge we develop along the way about doing so," says Di Franco. "I hope in the long run this empowers people working in small-scale settings to not just duplicate what can already be done in big labs and factories, but to actually innovate." ◗

Follow their work at openinsulin.org.

BUILD YOUR OWN

DIYBIO LAB

These tools will let you get started with some serious science

PATRIK D'HAESELEER
is a scientist, engineer, tinkerer, biohacker, and crazy ideas guy. He's the co-founder of Counter Culture Labs in Oakland, California.

Written by Patrik D'haeseleer

Autoclave

Incubator

Centrifuge

SO YOU'D LIKE TO TRY YOUR HAND AT SOME BIOHACKING. Professional labs cost hundreds of thousand of dollars to build from scratch, but you can get started for around $500 or less with a little improvisation and patience.

LARGE EQUIPMENT

Let's start with the pieces that will take up the biggest chunk of space in your lab: refrigerator, freezer, autoclave, and incubator. If you're just planning to run through an educational biotech kit at home, you can probably get away with clearing some space in your kitchen fridge. But if you *are* planning to do more than just a weekend project, invest in a dedicated fridge and freezer for your experiments. The vast majority of educational kits out there are perfectly safe, but the average novice might not realize that some classical experiments (like isolating unknown bacteria) can be hazardous to your health, especially if stored near food. Plus, you'll eventually need the extra space anyway, and it'll protect your experiments and edibles from cross-contamination.

Mini **fridges** can often be found on Craigslist for less than $50 — or free on Freecycle. If possible, avoid the kind that has a tiny freezer compartment at the top — they get iced over and are too small for practical use. Instead try to find a mini **freezer** that's the same size as your fridge or get a 2-door fridge/freezer combo. Many modern household freezers have an auto-defrost feature that briefly warms up the cooling coil a few times a day to keep ice from building up. The resulting temperature fluctuations can be detrimental to sensitive biological materials such as restriction enzymes. Check if there is a way to disable the auto-defrost circuitry or simply put your enzymes inside a styrofoam box in the freezer.

An **autoclave** is essentially a big pressure cooker that heats growth media or equipment above the boiling point of water

in order to sterilize them. Guess what — a regular pressure cooker makes for a fine autoclave as well. In a pinch, you can even use a microwave oven to sterilize growth media — just watch out for flash boiling! When you're ready to upgrade, professional autoclaves are surprisingly easy to find on Craigslist. Aim for one that's at least 7" in diameter so you can fit decent-sized flasks, and expect to pay a few hundred dollars.

An **incubator** is used to grow cells at a carefully controlled temperature. You might be able to score an old egg incubator or yogurt maker at your local thrift store. Or simply build your own by putting a heating pad with a thermostat in an old cooler. Pet stores sell heating pads intended to keep pet reptiles nice and toasty.

BENCHTOP TOOLS

A **centrifuge** is really useful for concentrating cells out of a liquid culture, separating DNA, proteins, and soluble components in complex mixtures, and more. Biohacker extraordinaire Cathal Garvey designed a 3D-printable "dremelfuge" that can be chucked into a Dremel power tool to turn it into a centrifuge. However, unless your 3D printer is well tuned they're prone to shattering at high speeds. You can order a high quality dremelfuge from Shapeways, but for that price you can also buy a cheap centrifuge on eBay. Inexpensive Chinese models running around 4000rpm can be had for $50–100, or 10,000rpm models for around $150. (Also see the how-to for a 3D-printable centrifuge on page 46).

A commercial gel **electrophoresis rig** can cost a few thousand dollars, but it's essentially just a DC power supply and a plastic box with two electrodes. You can make a dirt-cheap electrophoresis power supply using a hardware store dimmer switch and a bridge rectifier. The gel box itself can easily be made out of dollar store plastic containers and a stainless steel or platinum wire for the electrode (check out

the $21 gel box design by Cheapass Science at makezine.com/go/cheap-gel-box).

You may also want a **PCR (Polymerase Chain Reaction) machine**. Again, commercial gear can cost thousands of dollars, but there are several DIY designs available online, and OpenPCR even sells an Open Source Hardware PCR kit for $599. Like much biotech hardware, a lot of used gear from professional labs eventually winds up on eBay or Craigslist. The independent online biohacker store The Odin (the-odin.com) has been buying up inexpensive used PCR machines and reselling them after refurbishing.

You'll need some way to generate a sterile field in which you can manipulate microorganisms without fear of contamination. An open flame from an **alcohol lamp** or **Bunsen burner** will do for starters. A simple **laminar flow hood**, often used for mushroom cultivation or plant tissue culturing, blows ultrapure HEPA-filtered air over the cultures. Or if you want to get really serious, you could upgrade to a **biosafety cabinet**. Professional labs often leave their biosafety cabinets behind when they move, so if you cultivate the right contacts and are willing to put in some serious elbow grease, you may be able to score one for free.

LAB SUPPLIES

To manipulate small but very accurate amounts of liquids, you'll also need a set of adjustable **pipettes**. Cheap Chinese pipettes can be had from The Odin and a few other places for around $40 a piece — you'll want at least two or three different sizes to transfer anything from single-digit microliter droplets, to several milliliters.

Other smaller items you'll need include **digital thermometers**, a few small **digital scales** with a resolution of 0.01g or better, boxes of **nitrile gloves**, and a variety of **glassware** and **plasticware** (check your local dollar stores!).

Happy hacking! ⊘

Patrik D'haeseleer, Scott Covington/USFWS, SparkFun

Written by Quitterie Largeteau

BIOPRINTING PIONEERS

When it comes to fabricating with cells, these groups are leading the charge

Human cells cultured into a decellularized apple slice (left) and an apple carved into an ear shape (right) from Pelling Labs.

Bonnie Findley

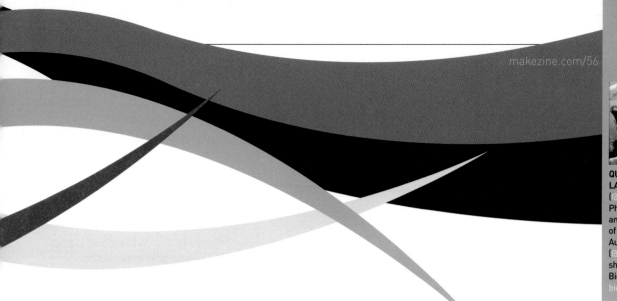

QUITTERIE LARGETEAU (@QuitterieL) is a Ph.D. in immunology and a communicator of sciences. With Aurélien Dailly (@dailylaurel), she cofounded Biohacking Safari: biohackingsafari.com

THE GREATEST BRIDGE BETWEEN THE WORLD OF MAKERS AND THE WORLD OF BIOHACKERS IS PROBABLY THE MIGHTY 3D PRINTER. The main difference is instead of using plastics, they're using biomaterials to build three-dimensional structures, and using special bioinks made of living cells to print messages and patterns.

HOW BIOCURIOUS STARTED BIOPRINTING

BioCurious is a mandatory stop among biohacker communities in North America.

This pioneering space, located in Sunnyvale, California, hosts a number of great people collaborating on the DIY BioPrinter project. Their bioprinting adventure started in 2012, when they had their first meetups. According to Patrik D'haeseleer, who is leading the project with Maria Chavez, they were looking for community projects that could bring new people into the space and let them quickly collaborate. None of the project leaders had a specific bioprinting application in mind, nor did they have previous knowledge on how to build this

kind of printer. Still, it appeared to be a fairly approachable technology that people could play with.

"You can just take a commercial inkjet printer. Take the inkjet cartridges and cut off the top essentially. Empty out the ink and put something else in there. Now you can start printing with that," D'haeseleer explains.

The BioCurious group started by printing on big coffee filters, substituting ink with arabinose, which is a natural plant sugar. Then they put the filter paper on top of a culture of E. coli bacteria genetically modified to produce a green fluorescent protein in the presence of arabinose. The cells started to glow exactly where arabinose was printed.

Modifying commercial printers for this, as they were doing, presented challenges. "You may need to reverse engineer the printer driver or disassemble the paper handling machinery in order to be able to do what you want," says D'haeseleer.

So the group decided to build their own bioprinter from scratch. Their second version uses stepper motors from CD drives, an inkjet cartridge as a print head, and an open source Arduino shield to drive it — a DIY bioprinter for just $150 that you can find on Instructables.

The next and still current challenge deals with the consistency of the ink. Commercial cartridges work with ink that is pretty watery. But bioink requires a more gel-like material with high viscosity. The DIY BioPrinter group has been experimenting with different syringe pump designs that could allow them to inject small amounts of viscous liquid through the "bio print head."

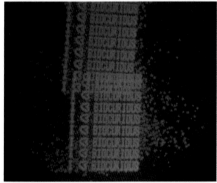

First major success with BioCurious' $150 DIY BioPrinter: fluorescent E. coli printed on agar with an inkjet printhead.

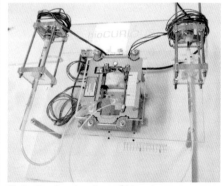

BioCurious' early printer: $11 syringe pumps mounted on a platform made from DVD drives.

Testing alginate as a promising DIY-able and accessible bioink at BioCurious.

Blob of hand-extruded layered alginate gel made at BioCurious.

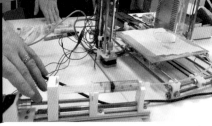

Converting a RepRap into BioCurious' latest 3D BioPrinter platform, with an Open Syringe Pump.

Closeup of a photopolymer print made with BUGGS' biocompatible resin.

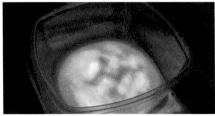

JuicyPrint uses G. hansenii and juice to make useful shapes from bacterial cellulose.

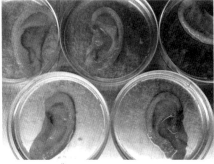

Pelling Lab's "Apple ears" during the decellularization process.

Counter Culture Labs' ghost heart only has connective tissue — all cellular material is removed.

MOVING TO 3D

Starting with an already existing 3D platform seemed like the best way to go beyond 2D patterns. The group first tried to modify their existing 3D printer by adding a bio print head directly on it. However, their commercial machine required some difficult reverse engineering and software modification to perfect the process. After a couple of months, this led to a dead end.

The RepRap family of 3D printers influenced the next step. After buying an affordable open source printer kit, the bioprinting team was able to switch out the plastic extruding print head for a print head with flexible tubes that connected to a set of stationary syringe pumps. It worked.

"The RepRap community is really what has made the whole 3D printing revolution possible," says D'haeseleer.

Soon enough there was a community around 3D bioprinting, tinkering at home and in biohackerspaces such as BioCurious, BUGSS, and Hackteria, all sharing their experiments.

WORKING WITH LIFE

The holy grail of bioprinting is generating 3D organs for transplants. Working with human or mammalian cells is complex. You need to have someone in the lab every day taking care of the cells and to keep everything as sterile as possible. Because of these obstacles, the BioPrinter group's current long-term project is to create a proof-of-concept functional plant organ and get it to photosynthesize. This will be an artificial leaf!

There hasn't been much work with plant cells, raising a lot of open avenues for research. You need to figure out what kind of cell types you will use, how to connect them together, what a 3D structure of a leaf looks like, etc. According to D'haeseleer, 3D printing with plant cells fits much better for a DIY community lab than actual mammalian cells.

Whether it works or not, the interest here is to test things and see how they grow. A commercial application is not the only purpose for biohackers, even though some scientists are a bit overwhelmed by the potential of their research.

"We are not very goal oriented, like we want to make a startup out of bioprinting and sell a product, make millions of dollars … There are not too many plants in desperate need of leaf transplants! We participate in this project because it's a fun thing to do. We make some progress week after week," says D'haeseleer.

3D BIOPRINTING WITH PLANT CELLS

When printing with plant cells, the first step is to figure out the material in which the cells are going to be held in place until they grow and make connections. Some current experiments at BioCurious use a gel-like material called alginate, which has very interesting properties. Sodium alginate is soluble in water, but viscous whereas calcium alginate solidifies instantly. It is similar to the spherification techniques seen in food science, where a solid droplet is full of liquid on the inside (See "Culinary Chemistry," page 78).

Several syringe pump designs are in testing now, all using the same comparison: one syringe pump contains the cells within an alginate solution, and the second contains calcium chlorite. When the two materials come in contact, the structure solidifies. Then you actually print a solid with embedded cells. Optimization is in progress.

Another challenge is deciding what cell type is needed. "Should we differentiate all the cells first and print the cells where we think they should go? Should we print undifferentiated cells and growth factors at the same time to let them differentiate and rearrange in situ?" The question is still open for D'haeseleer. The DIY group experimented with diverse cell types and did not recommend using carrot cells as people usually do. These stem cells are undifferentiated, which means they can give rise to different cell types under good conditions, but they are often contaminated.

Maria Chavez, BUGSS, Alasdair Allan, Andrew Pelling, Patrik D'haeseleer

OTHER GROUPS WORKING ON BIOPRINTING

BUGSS — BALTIMORE

Baltimore Underground Science Space is currently building a platform call 3DP.BIO that aims to connect scientists, engineers, and designers to accelerate research and development. They focus on resin printers, developing the control software along with a biocompatible resin that can be used to make 3D scaffolds for cell growth.

LONDON BIOHACKSPACE

The London Biohackspace's JuicyPrint machine prints using the Gluconacetobacter hansenii, a bacteria that is easy to grow using fruit juice as a food source. G. hansenii produces a layer of bacterial cellulose, a strong and exceptionally versatile biopolymer. However, the bacteria have been genetically modified to make them unable to produce cellulose under a light source. By shining different patterns of light onto successive layers of the culture, the structure of the final product can be manipulated, resulting in useful shapes made of cellulose.

PELLING LAB

Another way to grow tissues or organs would be to use an already existing 3D structure as a scaffold for cells. Andrew Pelling describes the process: "You slice an apple, wash it in soap and water, then sterilize it. What's left is a fine mesh of cellulose into which you can inject human cells — and they grow." His lab is now doing that to grow human ear prototypes.

COUNTER CULTURE LABS

Why 3D print when you can use already shaped forms? Case in point, the surprising example of a pig heart project from Oakland, California's Counter Culture Labs.

They make it by stripping out all the cells from a donor organ — a pig heart — leaving only the connective tissue to make it a "ghost" organ. Then, the idea is to repopulate it with the cells they want to grow. ⊘

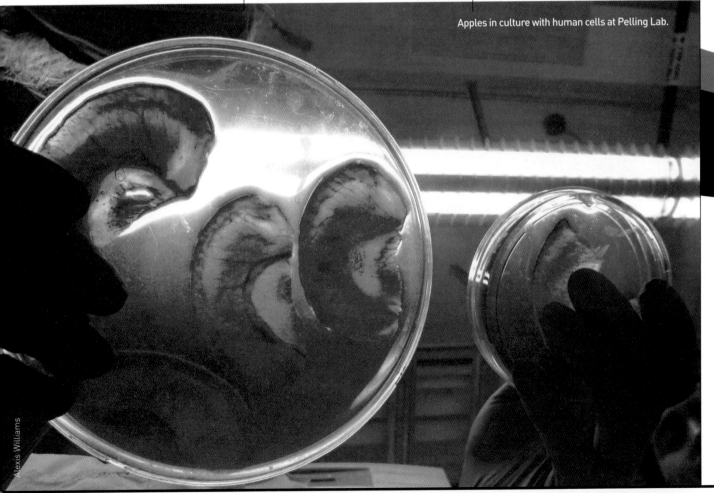

Apples in culture with human cells at Pelling Lab.

Alexis Williams

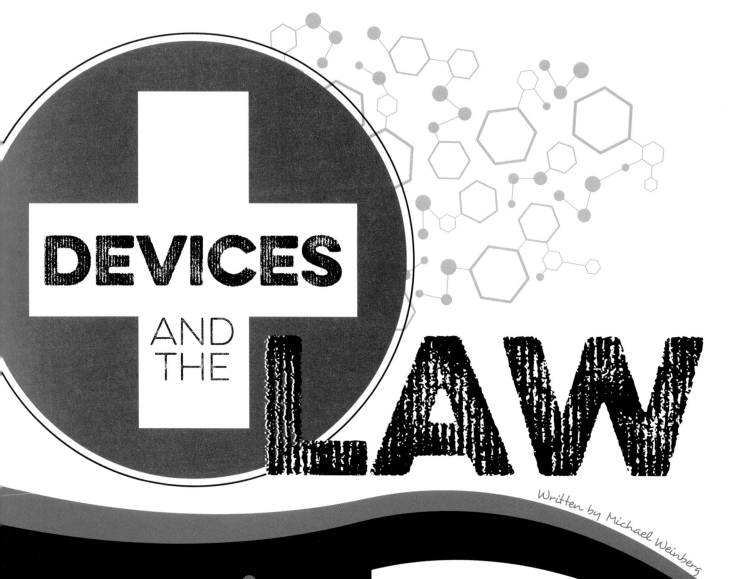

DEVICES AND THE LAW

Written by Michael Weinberg

The FDA recently published guidance to help distinguish between medical devices and wellness devices: makezine.com/go/medical-device-distinction. You can find additional resources in this article online at makezine.com/go/biohacking-policy.

MICHAEL WEINBERG In his spare time, Michael is the President of the Board of the Open Source Hardware Association (OSHWA) and the General Counsel of Shapeways. Despite that, nothing in this article is OSHWA's or Shapeways' fault, and it certainly isn't legal advice. You can find Michael online @MWeinberg2D and at michaelweinberg.org.

When does your project become a medical tool that the FDA wants to regulate?

IT'S ALL GOOD FUN TO 3D PRINT A DNA EXTRACTION CENTRIFUGE, make a homebrew heart rate monitor, or build an Arduino-powered bioreactor. But as you delve deeper into the world of biohacking, you may eventually wonder if (or when) the government will take an interest in what you're doing.

There are lots of good reasons for government regulation here. When you take a drug, get a device implanted in you, or have a medical test done, it can be comforting to know that someone is ensuring that the drug isn't poison, the device won't explode inside you, and the medical test won't insist that your ears are falling off (assuming that your ears aren't actually falling off).

THE SPIRIT OF '76

For all of their benefits, regulations can also prevent good things from happening. Processes can be expensive to comply with and testing can take time. Small-scale participants can have trouble identifying rules relevant to them, let alone comply with them. That makes regulation, like so many things in life, a balance between safety and doing new things.

Unfortunately for DIY biohackers, the last time Congress passed a major medical device law was in 1976. Back then, only large companies had the ability to manufacture medical devices. Congress didn't imagine that people would be making them at home. It certainly didn't imagine that the at-home inventors would be able to distribute them to the world over the internet — and possibly do so without any commercial infrastructure.

Nonetheless, the regulatory structure of 1976 remains in force today. Devices are sorted into three tiers based on their intended use. Class I devices are regulated under "general controls," which are generally light-touch rules designed to handle lower risk devices. Class II devices are subject to special controls because they are potentially more dangerous. Class III devices require

premarket approval before being distributed. (We're just setting the scene, but if you have specific questions about the three-tier system or about your specific device, you should speak with a lawyer.)

GENERAL WELLNESS VS. MEDICAL

For our purposes, the question is not how medical devices are regulated, but rather when your project becomes a medical device that the FDA wants to regulate. The difference is between a "general wellness" device, which isn't regulated, and a "medical" device, which is. Distinguishing between them involves two major factors.

The first is if the device is geared towards "general wellness." The more the device is oriented towards general wellness, the less likely it is to be regulated by the FDA. General wellness is a focus on overall health instead of specific diseases. Think about jogging — it's a healthy lifestyle choice that impacts a range of diseases, heart disease among them. But jogging isn't a specific treatment for heart disease in the same way that prescription medication is. General wellness devices tend to measure things like fitness, sleep, concentration, heart rates, or physical impact in order to help you live a healthier lifestyle.

The second factor is whether the device poses a low risk to safety. Not surprisingly, if it doesn't pose a safety risk, the FDA is more likely to classify it as a general wellness device. Conversely, if a device could pose a significant risk — even due to misuse — the FDA is likely to want to take a closer look. Helpfully, the FDA provides four characteristics to look for in a device, any one of which would suggest that the device poses a higher-than-low risk to safety:

- Is the product invasive (does it penetrate the skin)?
- Is the product implanted?
- Does the product use a technology that is itself risky if not regulated (think medical lasers or radioactive elements)?

- Are similar existing products regulated?

That last point is important. If you are working on a device and all of the established commercial devices in the space are regulated, the FDA will probably take a keen interest in your device as well.

What's the takeaway here? Keep hacking and building new things. But as you start building devices that are geared toward treating specific diseases or that can do real harm if they malfunction, keep in mind that the FDA might be interested. At that point it is probably time to put down this magazine and find yourself a lawyer. ●

Hep Svadja

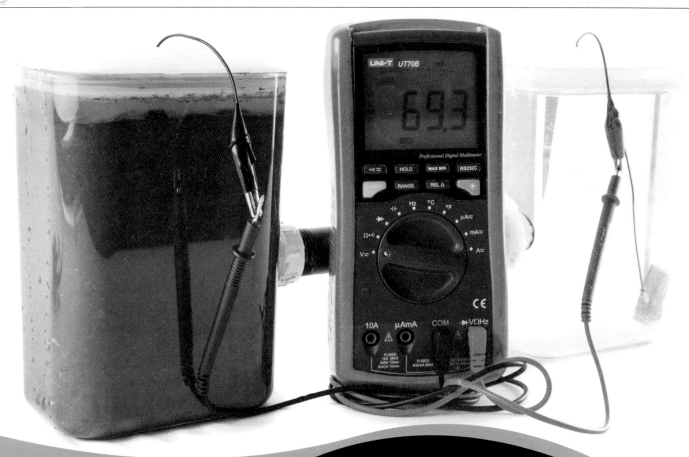

Written by Jendai E. Robinson

Hep Svadia

TIME REQUIRED: 1–2 HOURS   COST: $20–$40

MATERIALS

- » **Containers with lids, 1 quart, plastic (2)**
- » **Pond sludge or topsoil**
- » **Salt, 1 cup**
- » **Agar, about 6 grams** Agar is a jelly-like substance obtained from algae.
- » **Distilled water**
- » **Container with lid** to store mud
- » **PVC pipe nipple, ¾" × 6"**
- » **PVC pipe fittings, ½" slip × ¾" threaded (2)** to match the pipe
- » **Plastic wrap**
- » **Copper wire, insulated: red and black**
- » **Aluminum mesh, 4"×2" (2)** window screen or Phifer screen repair mesh
- » **Alligator clips (2)**
- » **Paper clips (optional)**

TOOLS

- » **Drill**
- » **Hot glue gun**
- » **Marker and ruler**
- » **Gloves**
- » **Multimeter**
- » **Wire stripper**
- » **Aquarium air pump (optional)** with air tubing

Create renewable, carbon-neutral electrical power from bacteria

MICROBIAL FUEL CELL

WITH SOME MUD, SALT, AND WATER, YOU CAN CREATE A CLOSED CIRCUIT THAT GENERATES A CURRENT. This is called a *microbial fuel cell (MFC)*, a device that uses bacteria to create electrical power by oxidizing simple compounds like glucose or organic matter in wastewater. Given the finite supply of fossil fuels, this biofuel cell is a promising approach for generating power in a renewable, carbon-neutral way.

The fuel cell works when bacteria attach to the electrode in an anode chamber of a cell that is oxygen-free. Since the bacteria don't have oxygen, they transfer their electrons to the anode instead. The cathode however is exposed to oxygen; thus, the two electrodes are at different potentials and create a bio-barrier or a "fuel-cell."

1. MAKE THE SALT BRIDGE

Prepare 1 cup of agar solution according to the instructions on the packaging. Add ½ cup of the salt to the water.

Cover one end of the PVC pipe nipple with plastic wrap to contain the prepared agar solution. Place the pipe vertically in a dish, then pour in the solution to fill, and allow it to cool (Figure Ⓐ).

2. OBTAIN A MUD SAMPLE

The good stuff comes from the *benthic zone* — the bottom of a body of water. This is where you'll find the electrochemically active anaerobic bacteria. If the sample is being collected from the bottom of a creek, pond, or lake, it should be black in color. Topsoil mixed with distilled water can also work if it contains enough anaerobic bacteria. Place the mud in a container and cover it.

3. BUILD THE HOUSING

Using a permanent marker, outline a hole on the side of one of your plastic containers large enough to fit the ½" side of a PVC fitting. With a ruler, measure the location of the mark and make a mark at exactly the same location on the side of the second plastic container. Ensure that your marked outlines are exactly opposite and facing each other, then cut out the holes.

Mark the center of each of the two lids for the containers. Drill a small hole in each lid for the copper wire to run through, and an optional hole on one of the containers for the placement of an air pump. Place the PVC fitting into the holes and glue into place. Allow glue to dry (Figure Ⓑ).

> ⚠️ CAUTION: Some adhesives can be irritating to the skin. I wear gloves while handling glue.

4. PREPARE THE ELECTRODES

Fold the aluminum mesh into two squares about 2"×2". Strip both ends of the red and black copper wires, and connect one end to a mesh square. Bind by folding the wire tight (Figure Ⓒ) or use a paper clip.

Insert the other ends into the pre-drilled holes on the containers and seal with glue.

5. ASSEMBLE THE FUEL CELL

Insert the PVC nipple into the threaded fittings on the containers and hand tighten.

Insert the air pump tube into the pre-drilled hole on the cathode container and seal with glue (optional).

Insert each end of the salt bridge into the pipe connectors and tighten until it is securely in place. Be sure to create a watertight seal (Figure Ⓓ).

6. FILL THE CONTAINERS

Using gloves, fill the first container halfway with your sludge. Take one of the electrodes and bury it. Remove any air bubbles and continue filling the container. This will be your anode.

Next, fill the second container with distilled water. Add the remaining ½ cup of salt and stir. This will be your cathode. Place your second electrode in the saltwater.

Seal both containers with their lids. Attach alligator clips to the ends of the protruding copper wires.

Optionally, turn on your aquarium pump to aerate the cathode solution. This allows for faster exchange of the electrons and consequently increases the voltage output.

USE IT

The performance of your biofuel cell can be evaluated by measuring its voltage output. Attach the respective ends of the alligator clips to your multimeter, then measure the voltage between the anode and cathode. You can expect around 0.2V depending on the amount of bacteria present.

Next, try attaching a resistor and measure your biofuel cell's power output in watts. Or connect an LED to the ends of the anode and cathode wires to see if there's enough energy to light it up.

Additionally, applying a small voltage to the bacteria produced at the anode can modify the cell. By not using oxygen at the cathode, you should be able to produce pure hydrogen gas. This modified process is known as a *microbial electrolysis cell (MEC)* and is based on the idea that fuel cells produce electricity whereas electrolysis produces hydrogen. ✏️

....

To see different biofuel cells in action, and to share your own, visit makezine.com/go/biofuel-cell.

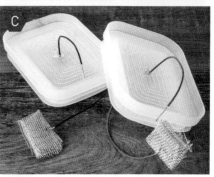

JENDAI E. ROBINSON works in the Center for Nanotechnology at NASA. She is currently a NASA Harriet G. Jenkins Fellow.

Sydney Palmer

3D-PRINTED CENTRIFUGE

Make this low-cost lab tool to extract DNA on your desktop

Written by ProgressTH

TIME REQUIRED: 2 DAYS   COST: $40-$60

MATERIALS

» 12V DC power supply, 2.1mm plug
» Female power jack, 2.1mm barrel
» Rocker switch, 2 pin on-off, 21×15mm 3A/250V
» Potentiometer, 10kΩ, linear taper aka B10K potentiometer, shaft length 10mm, shaft dia. 6mm, base dia. 18mm
» Arduino Nano
» Brushless motor, 12V, drone-specified, 1806/2400 with accompanying bolts and rotor nut
» Electronic speed controller (ESC) for drone motor
» Bolts, 22×3mm and compatible nuts (2)
» Bolts, 14×3mm and compatible nuts (4)
» Screws, 16×3mm (2)
» Jumper wires

TOOLS

» 3D printer and PLA filament
» Rotary tool with small collet (optional) or use super glue

BRIAN BERLETIC is co-founder of ProgressTH, a Bangkok-based design lab and media platform that focuses on using technology to solve real-world problems. They designed and built the equipment for F.Lab, a local DIYbio group, and have a growing collection of tools and equipment on their Thingiverse pages at thingiverse.com/ProgressTH/designs and thingiverse.com/F_Lab_TH/designs.

BIOTECHNOLOGY IS POWERFUL, BUT ONLY FOR THOSE WITH THE TOOLS to experiment with and utilize it. The DIYbio movement seeks to put the tools and techniques used in well-funded laboratories around the world into the hands of ordinary people who have an interest but not the means to investigate biology.

One of these tools is the centrifuge. Centrifuges come in many shapes and sizes to fit a wide variety of laboratory needs. There are large machines with precise controls for RPMs, G-force, timers of all kinds, and even ones with temperature control. Then there are mini-centrifuges used for simple DNA extraction and quick-spins for mixing the contents of test tubes.

This 3D-printed DIYbio mini-centrifuge was designed to do the latter and has actually been used in a real university biology lab doing real protocols. Building one is easy, and hopefully after you're done reading this, you will have ideas of how to improve on this one, or maybe the inspiration to tackle other types of otherwise inaccessible and expensive pieces of equipment with 3D printing.

PRINT THE PARTS

Go to F.Lab's Thingiverse page for the centrifuge (thingiverse.com/thing:1175393) to download the STL files. Because of the size of the parts, you may need to run multiple print jobs — this gives you a chance to switch colors like we did (Figure **A**). Print infill of 30% is recommended. Be sure to duplicate the feet so that you have 4 in total.

PROGRAM THE ARDUINO

It's a good idea to program your Arduino first and test everything out before assembling the entire centrifuge. Upload the code found on the project page (makezine.com/go/3dp-centrifuge) to your Arduino. Wire everything together as in the diagram (Figure **B**), but make sure to use only temporary connections between the 3 drone motor wires and the ESC, because you'll need to disconnect them and reattach them during the assembly process.

Hep Svadja

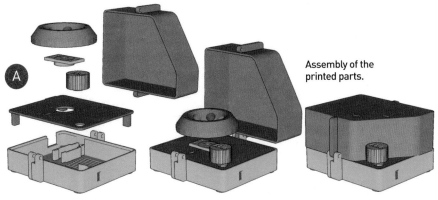

Assembly of the printed parts.

ASSEMBLE THE CENTRIFUGE

Attach the drone motor to the motor mount piece with its bolts. Then use the four 14×3mm bolts and nuts to attach the motor mount piece to the case cover (Figure Ⓒ).

Feed the motor's 3 wires through the case cover via the oval hole where they will then be connected to the ESC (Figure Ⓓ). It is now possible to wire the rest of the electronics together. Be patient and follow the diagram carefully.

Once wired up, gently force the rocker switch into the notch on the back, left side of the case. Install the female adapter similarly. Attach the wired-up potentiometer to the case cover, and then fit it into place into the space provided for it in the front, right side of the case.

With your Arduino Nano also wired up, you can either tuck it away inside (Figure Ⓔ), or you can glue or friction weld a piece of plastic behind it to hold it in place next to the mini-USB terminal hole on the right side of the case.

This centrifuge is still a prototype, so carefully pack the wires in but don't worry if it looks like a mess. Just make sure you didn't leave exposed wires or contacts, and that nothing is being pinched or crushed when you put the cover on.

Using the feet as washers, attach the case cover to the case body with the 3 screws, and put the knob onto the potentiometer shaft with a thin piece of cardboard inside to help it stay on if necessary. Add the rotor by simply placing it over the motor's shaft and then tighten it down with the nut. It should be tight, but don't go overboard.

Finally, add the 22×3mm bolts and nuts to attach the lid on the back of the case, and to use as a fastener in front when transporting the centrifuge (Figure Ⓕ).

USING YOUR CENTRIFUGE

The centrifuge is a little tricky to use — the ESC interprets everything you try to do as if it is still attached to a very expensive drone.

Sydney Palmer

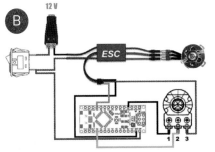

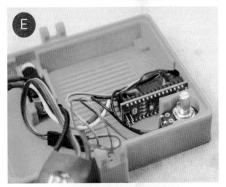

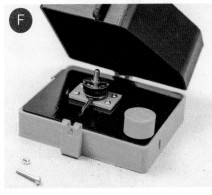

NOTE: People have pointed out you can use a servo tester board instead of the Arduino and ESC. We haven't tried this, but it may be less tricky for beginners. You may have to modify the 3D model — we include the SketchUp files for this exact purpose.

Too much power, or fluctuations in power, can cause the motor to stop working and the ESC to reset. Some tips and notes:

- When you turn on your centrifuge using the rocker switch, the ESC will beep, then calibrate itself.
- Slowly turn the knob until it reaches halfway and begins spinning. Turn it the rest of the way to get it to spin faster.
- We recommend that you only spin the rotor at 1–2 minute intervals. We use a phone app stopwatch to time it. **Always operate it with the lid down**.
- Be sure to **always balance the rotor**. If you're using only one test tube, make sure to place another opposite it, even if it is empty. The rotor must always be balanced or you risk subjecting it to uneven and possibly destructive forces.

GOING FURTHER

In addition to the centrifuge, we also have developed a magnetic stirrer as well as an electrophoresis system. Together with our tube rack blocks, this offers everything you need to do basic DNA extraction and analysis. ⊘

Get the code and more details at makezine. com/go/3dp-centrifuge.

DNA

Extract
DNA from
strawberries
in this science
experiment
you can drink

Written by Bonnie Barrilleaux

DOUBLE HELIX DAIQUIRIS

TIME REQUIRED: 30–45 MINUTES **COST:** ABOUT $40

MATERIALS
» **High-proof alcohol (greater than 80 proof), ice cold** such as Bacardi 151, Don Q 151, or Everclear
» **Strawberries, frozen and sliced**
» **Pineapple juice concentrate, frozen**

TOOLS
» **Zip-lock bags**
» **Strainer**
» **Bowls**
» **Ice**
» **Scale or measuring cups**
» **Narrow glasses for serving**

BONNIE BARRILLEAUX is a staff data scientist at LinkedIn. She has a Ph.D. in chemical engineering, experience in genomics research, and an insatiable urge to mix science and art.

EXTRACTING DNA FROM SPIT OR FRUIT IS A FAVORITE SCIENCE FAIR/MAKER FAIRE DEMO. This is the adult version of that, inspired by Mac Cowell's 5 minute DNA Extraction in a Shot Glass on Instructables. Especially his final picture, where he drinks his own salt-detergent-rum spit cocktail. Hey, if you're going to use high-proof rum, you might as well make it tasty, right?

Extracting DNA from plant or animal sources is a fairly straightforward procedure, and can even be done at home. However, to our knowledge, no one had previously created a DNA extraction protocol that also functions as a cocktail recipe. The DNAquiri is an attempt to fill this scientific void. The cocktail consists of a strawberry puree layer and an alcohol layer, where the DNA from the strawberries is extracted into the alcohol layer. Because DNA is an extremely long polymer, when it clumps together in the alcohol layer it forms long strands that are visible to the naked eye and can be picked up with a toothpick.

1. MEASURE OUT INGREDIENTS
Put 250g (about 2 cups) strawberries and 75g (about ¼ cup) pineapple juice concentrate (Figure Ⓐ) into a zip-lock bag and seal it. Gently crush the strawberries

Hep Svadja

...QUIRI VS.
...DARD
...E DNA
...ACTION

...nts differ significantly
...rials for a standard home
...n. Key changes:

...rfactants such as dish
...d to lyse cells. These
...tasty. Instead, we start
...strawberries, in which
...haw process has already
...ells. We found that no
...as necessary when
...frozen fruit. Strawberries
...gave us the most DNA by far,
probably because commercial
strawberry strains tend to be
octoploid, i.e., they have 8 copies of
DNA per cell.

● Salt is usually added to enhance
precipitation of the DNA. We found
salt unnecessary and were able
to obtain reasonable yields while
omitting it.

● Meat tenderizer is often added to
help break down proteins and free
the DNA; this mimics the effect
of proteases that would be used
in a lab setting. Instead, we used
pineapple juice, which contains the
protease bromelain. Note that canned
pineapple should not be used, since
the heat used during the canning
process deactivates bromelain.

● High-proof alcohol (greater than
80 proof) is required for DNA
precipitation. Since a daiquiri would
traditionally be made with rum, we
selected Bacardi 151.

little bit chunky.

4. SERVE

Put about 50ml of fruit pulp into a glass,
then gently layer 10ml of ice cold Bacardi
151 on top (Figure Ⓔ — we poured extra
for photography purposes). The strawberry
DNA will be extracted into the alcohol layer.
Swirling the glass gently may help promote
DNA moving into the alcohol phase. If you
prepare the cocktails in advance, don't wait
more than 15 minutes before serving. You'll
extract more DNA as time goes on, but the
DNA will start degrading after that.

Add a tiny umbrella; DNA can be collected
onto the end of the umbrella's toothpick by
twirling it, if desired (Figure Ⓕ).

RESULTS AND CONCLUSIONS

During the protocol development process,
we received comments on early versions
such as "gross," "salty," and "tastes like
soap." We are pleased to report that
feedback on the final product was generally
positive. The DNA looks disconcertingly
snotty when freshly extracted, but the texture
was not really noticeable when drinking.

You'll have a layer of very high proof rum
on top of you cocktail, so before drinking —
or even sipping — you'll probably want to
mix the alcohol into the fruit puree.

Tasters enjoyed the ability to scoop up
their drink's DNA on the end of a toothpick
and found the whole experience rather
intoxicating. Or maybe it was just the rum. ◓

Share your DNA extracting projects at
makezine.com/go/dna-daiquiris.

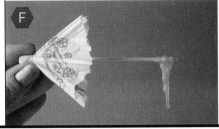

Skill Builder

TIPS AND TRICKS TO HELP EXPERTS AND AMATEURS ALIKE

Make Your Mark

Customize your gear with these simple secrets to screen printing

Written by
Tim Deagan

Screen frames are built
to handle heavy usage.

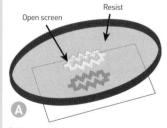

Only areas where ink should pass
are left unblocked.

"Pulling" a print refers to forcing
the ink through the screen with
the squeegee.

Halftone grays are possible with different
size black dots.

Halftone colors trick the
eye into mixing colors.

THE ENTHUSIASM OF THE OPEN SOURCE AND MAKER MOVEMENTS HAS LED TO QUITE A DEMOCRATIZATION OF TOOLS. Capabilities once reserved for professional or academic users have become staples of our daily lives. One of the earliest tool sets that changed the world after it became democratized was printing. This happened in the 15th century and again in the 20th. Inkjet printers are so cheap that they frequently cost less than the refills of their ink. The materials they print on are also varied and easy to acquire. T-shirt transfers, stickers, business cards, transparencies, and other media are common materials for personal computer users.

Yet computer printing's predecessors remain in use. The market for printing presses may have receded, but it hasn't disappeared. Of the many printing processes that continue to have utility, screen printing manages to remain a viable practice among professional and DIY users. T-shirts may be the most popular choice, but there are many other uses for this venerable technique. The basic activity is dead simple and can be done with surprisingly primitive equipment. As with many interesting skills, incredible results can be achieved by learning the subtleties of the art.

HOW DOES IT WORK?

The fundamental concept of screen printing is to stretch some kind of tightly woven material and block specific parts of the material so that when ink (or paint) is forced through it, the ink only passes through the unblocked portions. The blocking, or **resist**, creates a negative image on the screen of what you want on the printed surface (Figure Ⓐ). Screen printing is often called "silk screening" because silk was once the best available material. Synthetic materials have largely replaced silk over the years, but the name remains.

The screen maintains its stretch by attachment to a frame of some sort. Many screens are stapled or glued to a wooden frame, but there are lots of ways to use metal or even plastic frames. Using a frame that won't warp or distort the screen is important for reuse (see the well-loved screens on previous page).

Most screen printing techniques use a squeegee to evenly force the printing medium through the screen onto the print surface. This operation is referred to as **pulling a screen** since the squeegee is usually drawn toward the printer for the print stroke (Figure Ⓑ).

Screen printing is a binary operation; if you're printing with black ink, you cannot directly print gray. However, special techniques have evolved to emulate fades and gradations. Instead of mixing white with black to get gray, a **halftone image** is created that uses proportionally sized dots to create the perception of gray, or pink from red and so forth (Figure Ⓒ). This method also allows for mixing colors on the print. A 50% halftone of yellow printed on top of a 50% halftone of red will crea the perception of orange (Figure Ⓓ).

Multicolor prints require a screen fo each color (except for some interestin effects achieved by mixing multiple c of ink or paint on the screen). A print shop that prints color images will usu ally have a screen press with four scr arranged in a manner that allows th print surface to be carefully register on each screen until each of the col (cyan, magenta, yellow, and black a common) has been pulled (Figure

For this skill builder, we'll stick t ing a single color without halftones is a wealth of opportunities for this ap proach; printing labels on control surfaces, adding logos to products, making basic T-shirts, or even putting a resist layer onto a surface for etching printed circuit boards.

MAKING THE SCREEN

Screens can be created with a variety of methods. You can paint the resist material onto the screen by hand, you can use a vinyl cutter to create a stencil and apply it directly to the screen, or you can use

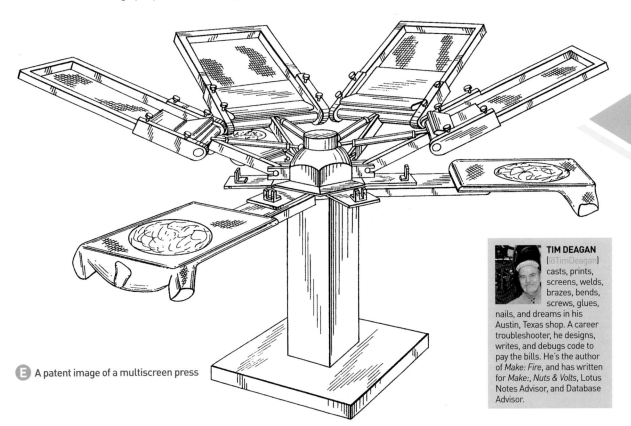

Ⓔ A patent image of a multiscreen press

TIM DEAGAN
(@TimDeagan) casts, prints, screens, welds, brazes, bends, screws, glues, nails, and dreams in his Austin, Texas shop. A career troubleshooter, he designs, writes, and debugs code to pay the bills. He's the author of *Make: Fire*, and has written for *Make:*, *Nuts & Volts*, Lotus Notes Advisor, and Database Advisor.

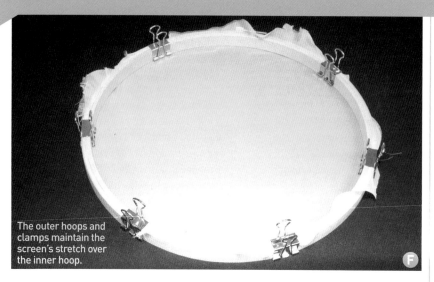

The outer hoops and clamps maintain the screen's stretch over the inner hoop.

Trace the outline of the image onto the screen.

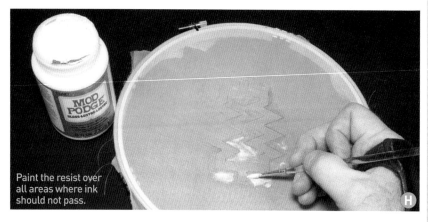

Paint the resist over all areas where ink should not pass.

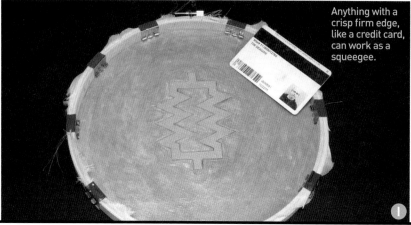

Anything with a crisp firm edge, like a credit card, can work as a squeegee.

photo emulsion to transfer a transparency onto the screen. Each of these methods has pros and cons. We'll do the simplest for this first screen and you can let your level of interest determine which other methods you pursue.

Most art supply stores carry basic materials for screen printing and online sources have everything you need whether you're printing for fun or running a screen printing business. In the interest of simplicity, we're going to use commonly available materials to demonstrate the ease of basic printing. Feel free to upgrade if you're interested.

Rather than going through the process of stretching screen material to a wooden frame, we'll use an embroidery hoop for holding our stretched screen material. In the spirit of low budget skill hacking, we can use any sheer polyester material or organza for our screen instead of commercial screen cloth. I bought a bathroom curtain at Goodwill. The results won't be as good as using a high thread count fabric made specifically for screen printing, but it will still work. Cut the material so that you have a section slightly larger than the hoop. Lay the material across the smaller hoop ring, then put the larger ring around it and tighten while stretching the material as tightly as possible. I used small binder clips to lock the cloth in place (Figure F).

ADD YOUR DESIGN

Next we need to create the art. We'll hand paint it onto the screen, so you'll want to match the art to your dexterity level. For your first effort, avoid fine lines and stick to something basic. Print the art out on paper to the same size you want to screen it. You'll need a couple inches around the art to move the squeegee, so don't make the art bigger than half the size of the hoop.

Tape the print of the image to a solid surface so it won't move and place the screen on top of it. Use a soft pencil to trace the outline of the image onto the top of the screen (Figure G). We're going to use Mod Podge for resist. Commercial resist products are available that are better for serious efforts, but that's true for pretty much every aspect of this project.

Our goal is to block passage of the ink through any tiny holes in the screen other than the ones we intentionally left uncovered. We will paint the Mod Podge resist onto the screen to cover areas where we don't want paint to go through. It may take multiple coats of resist to fully block the cloth (Figure H).

Tim Deagan

GATHER YOUR TOOLS

Squeegees are traditionally rubber with a firm but flexible edge. They come in a variety of sizes, but it's important that the squeegee be wider than the art so that you can pull the print in one motion. We can use a credit or club card (Figure I) as our impromptu squeegee, but anything with a firm plastic edge will work.

For ink, we'll use acrylic paints. Commercial inks are available that are designed specifically for different surfaces. Screen printing ink (or paint) has to be thick enough so that it doesn't run when you push it through the screen.

GET READY, GET SET, PRINT!

Whatever your intended print surface will be, it's always a good idea to pull a couple of prints onto paper or cardboard to get a feel for the squeegee, ink, and screen. Place some protective newsprint onto your work area and lay down a piece of cardboard. If things are likely to slide, tape everything in place. To avoid blurring, it's very important that neither the material being printed nor the screen move while printing.

Pulling the print is typically done in three motions; first drop some ink onto the screen, use a good dollop so that you'll have more than enough for a print (Figure J). Second, use the squeegee without pressure to lightly spread an even coating of ink over the art (Figure K). Third, press the squeegee onto the screen above the art, hold it toward you at 45° and pull it toward you while pressing evenly, but not too firmly, down, forcing the ink through the screen (Figure L). Learning the correct pressure is an important reason to run through some test prints first.

Lift the screen directly off of the print. Commercial (and many home) screen printing rigs mount the screen on a hinge so that it can be easily lifted up. Give it time to dry before touching it (Figure M). If you're printing fabric ink on clothing, you'll need to heat-set the ink with a heat gun before washing.

We've really only scratched the surface of screen printing. For another great project, try cutting stencils with adhesive-backed vinyl that you apply to the screen (use a heat gun to get the vinyl to stick). Photo emulsions allow you to make screens with incredibly fine detail. Screen printing is a great way to get professional-looking patterns onto your projects or to do short runs of clothes, posters, or art. I hope you'll keep looking into the next steps now that you've seen how easy it is! ◑

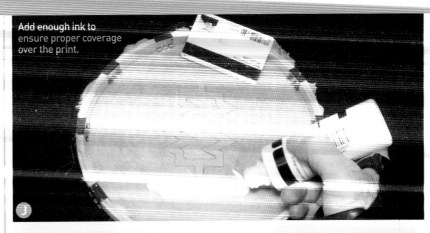

Add enough ink to ensure proper coverage over the print.

J

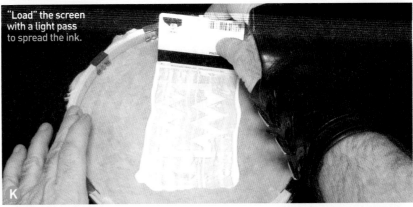

"Load" the screen with a light pass to spread the ink.

K

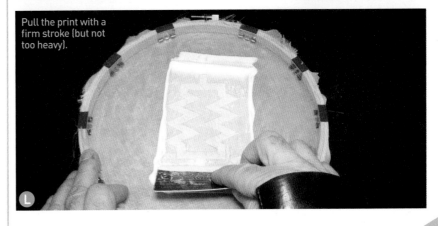

Pull the print with a firm stroke (but not too heavy).

L

You can print on anything your screen will lay flush against.

M

CALEB KRAFT is Senior Editor at *Make:*. While he is always up for a good experiment, proper documentation eludes him.

Learn the Lingo:
Machining Metal *Written by Caleb Kraft*

Get literate in machine-shop speak with this primer

WHEN YOU'RE JUST JUMPING INTO A NEW SKILL, THE LINGO THAT THE EXPERTS USE CAN BE CONFUSING and have meanings you didn't anticipate. In order to help bring you up to speed on the terms you'll need to understand when discussing a specific skill, we launched a new series called *Learn the Lingo*. You can see the installments at makezine.com/go/learn-the-lingo.

In this issue, we're going to explore the vocabulary of **machining**. Specifically, we're talking about the process of cutting away bits of metal to make shapes. This can be done with hand-based tools called mills or lathes. Though you'll hear about mills and lathes in woodworking, the term machining tells us that we're going to be talking about metal.

"People definitely get confused about terminology when machining, which can be quite a big deal when you're dealing with high-precision, spinning metal cutters," says Chris Fox from Tormach, makers of prosumer-grade CNC milling machines. "This can be blamed on the fact that like machining itself, where there are generally a bunch of different ways to do the same process, several different machining terms can often identify the same thing or slightly nuanced versions of the same thing. The key to learning machining is to learn by doing, but it helps to have a mentor or guru (YouTube is a great place for this) to keep your lingo on the right path."

LINGO TO LEARN FOR YOUR FIRST DAY MACHINING METAL

» **Feeds and Speeds** – A calculation to determine the best way to cut through a piece

Tormach

of metal. These need to be adjusted according to the hardness of the material, the surface finish you are looking for, and what type of cut you are making.

Workholding – An apparatus or structure designed to hold the material you are cutting with your CNC machine. On a mill, this is most often attached to the table, whereas on a lathe, this is most often the element that is spinning.

» **Toolholding** – The structure or apparatus designed to hold your cutting bit or tools. On a mill, this is often the element that spins, whereas on a lathe, this is the element that is secured to your table.

» **Fixturing** – Similar to workholding, this is the apparatus or structure that holds your work, but which is often custom-made to hold your specific part.

» **CAD** – Computer aided design. CAD programs allow you to digitally create a part that can be milled, 3D printed, or rendered.

» **CAM** – Computer aided machining, sometimes known as computer aided manufacturing. CAM is a computer program that takes a CAD file (IGES, STEP, etc.) and allows you to create tool paths for the cutting process on a CNC machine. CAM tells your CNC machine where to go, how fast to move and spin.

» **G-code** – This is the coding language that a CNC controller reads to tell the machine what movements to make and the speeds at which to move, among many other functions.

» **Post Processor** – While most CNC controllers run on standard forms of G-code, every controller has a different dialect of control code. A post processor acts as the translator to get a CAM file into the proper code that can be read by a specific machine.

» **Chatter** – A noise that can occur during cutting, which can be caused by an array of different factors, including (but not limited to) dull tools, incorrect speeds and feeds, and too much tool stick-out. Too much chatter mostly affects the finish on a part, but it can also be signs of inefficient cutting, looming tool breakage, or improper workholding.

» **Depth of Cut** – Closely related to speeds and feeds, depth of cut is how much material is being removed with each pass of the tool.

This is another adjustment that can be made to improve surface finish and/or the life of your tools.

» **Thou** – A unit of measurement equaling 0.001" or one thousandth of an inch.

» **Tenth** – A unit of measurement equaling 0.0001" or one ten-thousandth of an inch. **Not to be mistaken with 0.1"** or one tenth of an inch.

» **Conversational** – An interface for machining that is done at the controller; a method of CNC machining without CAD or CAM, for very simple operations.

» **Work Coordinate System** – Also known as WCS, this tells the CNC machine where a part is located relative to the machine's home.

» **Concentricity** – When two circles of different sizes share a common center, they are concentric.

» **Runout** – The concentricity of a tool in relation to the spindle that is spinning it. More runout means that a tool will be less accurate in the cutting process — a tool with high runout has a harder time staying true to the programmed tool paths.

» **(Tool) Offsets** – Tells the CNC machine where the tip of the tool is relative to a known surface (like the nose of the spindle). ◑

GOING FURTHER

- Advance CNC Machining has an exhaustive glossary at advancecncmachining.com/machining-glossary.

- CNC Cookbook has a similarly amazing CNC Dictionary at cnccookbook.com/CCDictionary.htm.

- Micro-Matics offers a glossary at micro-matics.com/cnc_swiss_glossary.html.

- If you still haven't found what you're looking for, check out All Words' Glossary of Machinist's Terms at allwords.com/machining-glossary-164-594.php.

Do you know any other helpful machining terms for beginners that we missed? Share them in the comments at makezine.com/go/machining-metal.

JORDAN BUNKER is a polymathic jack-of-all-trades who enjoys manipulating ideas, atoms, and bits. Find him in his basement workshop in Oakland, Calif.

Fabricating Fasteners

Written by Jordan Bunker

Learn how to create your own threads with taps and dies

Taper

Plug

Bottoming

SCREWS AND BOLTS ARE SIMPLE OPTIONS FOR FASTENING TWO PIECES OF MATERIAL TOGETHER. As you might have guessed, the secret is in the threads, but how would you go about making your own? In this Skill Builder, we'll go over the tap and die, a set of tools that allows you to make your own threads wherever you need them.

TAPS

A **tap** is a tool that cuts **threads** into a hole so that a bolt can be screwed into it. While they look similar to bolts, taps are typically made of high-speed steel and have long channels ground into the sides, leaving gaps in the threads. As the tap turns into a hole, these channels allow the chips that are carved out of the material to break free and be ejected.

» SELECTING A TAP

Taps come in three main styles:

Taper — for large holes or harder materials. They start narrow and taper to full thread width. This means it's easier to begin threading, but they need to be turned farther in order to form a full thread in a hole.

Plug — the most common tap for general purposes. It has a slight taper, but allows for threading almost to the bottom of a blind hole.

Hep Svadja

Bottoming — for forming threads in the full length of a hole that has a bottom. It's recommended to use a taper or plug tap to form the initial threads.

» THREAD COUNTS

For every bolt, there's a corresponding tap with a matching diameter and number of threads per inch. Make sure that you choose the right one! Similarly, for each tap, there is a corresponding drill bit size that you should use to drill the initial hole. These relationships are typically printed on the packaging for the tap, or they can be found through a quick internet search.

If you're not sure what thread count a bolt is, you can use a screw-pitch gauge (Figure A) to match one of its blades to the thread profile.

QUICK TIP If you find yourself using taps often, it's a good idea to keep each tap with its corresponding drill bit. I keep duplicate drill bits just for my taps. That way, when I need to put a ¼-20 bolt somewhere, my tools are ready to go.

» SECURE THE PIECE TO BE TAPPED

This is one of the most important parts. Taps are very brittle, so if your tap or your material moves too much, it will break the tap off in the hole. To avoid that, it's a good idea to use clamps and align the hole so that it's either directly below you, or directly in front of you. This will help the tap stay true when you're turning it into the hole.

» SELECT THE TAP WRENCH

Since taps are just bits, they need a tool to turn them. The trusty tap wrench is specifically made for this (Figure B). For tapping holes ¼" or less in diameter, a 5"–7" tap wrench will do, but for anything larger, you'll want a longer handle. Insert the square end of the tap into the adjustable jaws of the tap wrench, then twist the handle that tightens the jaws.

If you don't have enough clearance for the handle, you can use a T-wrench (Figure C). These smaller alternatives are great anywhere a normal tap wrench won't fit.

» TAPPING PROCESS

Here comes the tricky part. Line up the tap with the hole, making sure that it's completely perpendicular to the material (it can be helpful to have a friend sight it from the side while you do this). Apply slight pressure and begin to turn the wrench (Figure D). You should feel the cutting threads begin to catch, so double-check that the tap is straight, and do not force the tap or it may break. After you've made a couple full revolutions into the material, you can stop applying pressure, and give the tap a small turn backward after each complete revolution to break the chips off and clean out the thread. Once you're done, turn the tap out, blow the chips out of the hole with compressed air, and test it with a bolt!

DIES

A **die** (Figure E) is basically the inverse of a tap: It cuts threads into the outside diameter of a rod. The rod's diameter will dictate which die you use.

» PREPARE THE ROD

To prepare the rod for threading, just file or grind a bevel onto the end so that it's easier to turn the die onto it (Figure F).

» PLACE THE DIE IN DIE STOCK

After you've selected your die, you'll need to put the die into the die stock, which is similar to a tap wrench, but has a socket that the die fits into. Most die stocks have a set screw on the side, which is lined up with the dimple on the die, and screwed in to tighten the die into the stock.

» CLAMP THE ROD IN PLACE

Rather than try and spin the rod into the die, we'll spin the die onto the rod. That means that the rod will have to be held securely, ideally in a vise. This also helps keep the rod straight while spinning the die onto the end.

» CUT THE THREAD

Cutting the thread is almost identical to tapping. Make sure that the die is aligned with the rod and apply pressure, then turn it onto the rod. Once the initial threads have been formed, and the die has "caught," you can remove pressure and continue turning (Figure G), remembering to give it a small backward turn once every revolution to break the chips. Once you've threaded the rod to the desired length, give it a test! ⊘

TIP If you're tapping a soft material like aluminum, brass, or cast iron, you won't need lubrication, but if you're tapping steel, some cutting oil will help you out.

Recalling
an Era

Written by Bob Murphy

Build a functioning cellphone that looks like it
was made by Motorola in 1940

Time Required: 16 Hours
Cost: $150

MATERIALS

- » **Feather Fona microcontroller** Adafruit #3027 adafruit.com
- » **Micro SIM Card** I used one from T-Mobile, special plan: $10 USD for the card, $3/month for 30 minutes text/talk. (Note: The T-Mobile Pay-as-you-go plan doesn't offer Caller-ID; you can't ID SMS senders.)
- » **Battery, 1200mAh LiPo** Adafruit #258
- » **Antenna, stick-on** Adafruit #1991
- » **NeoPixel Jewel LED** Adafruit #2858
- » **LED sequins (2) (optional)** Adafruit #1758
- » **Phone-style 3×4 matrix keypad** Adafruit #1824
- » **OLED display, monochrome, 0.96", 128×64** Adafruit #326
- » **Perma-proto PCB, small** Adafruit #1214
- » **Electret microphone, wired** Adafruit #1935
- » **Speaker, mini-metal with wires, 8 ohm, 0.5W** Adafruit #1890
- » **Battery extension, JST-PH, 500mm** Adafruit #1131
- » **On-off switch, pushbutton, 12mm** Adafruit #1683
- » **Wire, 22AWG and 26AWG, solid**
- » **Breadboard** for testing
- » **Headers, right angle**
- » **Headers, straight**
- » **Jumper wires**
- » **Screws, brass, #4/40, ¾", round head, slotted (4)**
- » **Threaded inserts, brass #4/40, ⅜" (4)**
- » **Wood, 12"×12"×½"** I used walnut.
- » **Bezel (optional)** for the back badge. Comes 2 per package from Bead Landing: makezine.com/go/brass-bezels
- » **Speaker cover material, vintage**
- » **Plastic, ⅛" acrylic, transparent yellow and green, 8"×8"** I used yellow for the speaker bezel and green for the microphone bezel.
- » **Plastic, light panel, smooth matte** to diffuse light behind the speaker bezel. Tap Plastics will cut to size (in-store only). Or improvise — card stock might work.
- » **Code and design files** for programming Fona and CNC and laser cutting: github.com/thisoldgeek/DieselPunk-Cellphone

TOOLS

- » **Soldering iron and solder**
- » **Desoldering tool**
- » **Multimeter** for checking connection continuity
- » **CNC router or mill with ⅛" down-spiral dual flute mill**
- » **Laser cutter**
- » **Pliers, needlenose**
- » **Diagonal cutters**
- » **Rotary cutting tool**
- » **Helping hands and/or Panavise**
- » **Tweezers, fine-tipped**
- » **Electrical tape, vinyl**
- » **Masking tape**
- » **Super glue**
- » **Hot glue gun**
- » **Epoxy**
- » **Arduino IDE software**

A Early concept sketch for Dieselpunk Cellphone

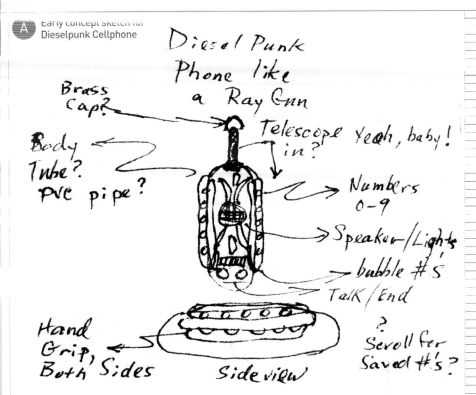

B

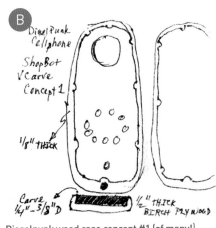

Dieselpunk wood case concept #1 (of many!)

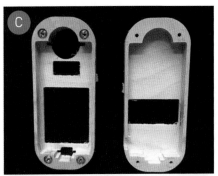

C

IT'S A GOLDEN TIME FOR MAKERS.

With a handful of components and access to common maker tools, it's now possible to build your own functioning cellphone that makes and receives calls and SMS texts, and even plays FM radio.

Adafruit's Fona microcontroller, with a GSM phone module, came out around the time I had just discovered dieselpunk (like steampunk, only the era from roughly 1930s to the end of WWII), and I was inspired to make a cellphone in an imagined retrofuture style. Something fun and artsy that actually made you think about our relationship to tech and culture.

I started out just drawing sketches on napkins at the coffee shop. I'm no artist, but these little sketches helped me visualize what I wanted.

Most of the concepts I came up with were way too advanced for me. I would have to improve my 3D printing skills (from rank beginner) to make the sketch shown in Figure **A**.

After a lot of thought and some prototypes (Figures **B** and **C**), I inched closer to the final concept.

I kept refining things and finally made a critical design decision. Up until late 2016, I had been trying to "re-invent the wheel" and create custom parts and features that

BOB MURPHY is an old retired guy who has plenty of time to steal other people's work and mash it into an unrecognizable hodgepodge that sometimes works.

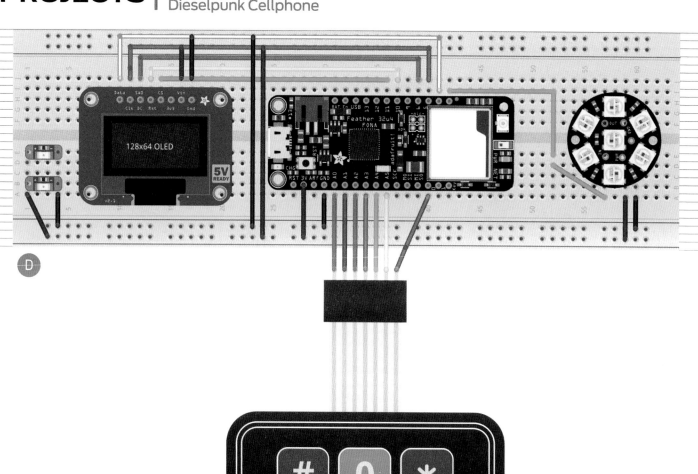

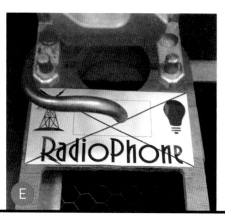

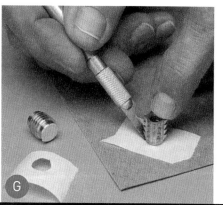

Fritzing, Bob Murphy

already had available solutions. Gee, *why not use off-the-shelf parts?*

Finally, I settled on a case I could make that could fit all the off-the-shelf components.

There were still design constraints. I wanted this cellphone to have about the footprint of an iPhone 6, but of course, it would be thicker, about one inch. Then Adafruit released its Feather Fona update. This was a great improvement over what I was doing, making everything compact: processor, Fona module *and battery charger* on one board! I had to modify my design a bit to fit the Feather Fona in — the case had to be made wider and I had to move a cutout for the USB charging port, among other changes.

At last, I had a case that achieved my goals, given some compromises for what I could do. Here's how I put the whole thing together.

KEY LEARNING: Starting with the case is exactly the opposite of what you should do! I should have begun with the components in the first place, gotten them to work, and then figured out how to make a case for them.

For complete details on the build, see my blog post at thisoldgeek.blogspot.com. You can get the code and CAD design files from github.com/thisoldgeek/DieselPunk-Cellphone.

DOWNLOAD SKETCH AND ASSEMBLE COMPONENTS
Download the Arduino sketch from the Github repository and upload it to the Fona. Assemble the components (Figure **D**) and make sure everything works.

CUT THE CASE AND BEZELS
Mill the case from ½" wood using the design files (with or without the back badge bezel) on the project's Github page. Sand the case and remove the holding tabs.

Laser-etch the front of the top case. I created a logo and graphics that are available on the Github page. You'll have to center the art on the OLED cutout for this (Figure **E**).

Cut out the speaker and microphone bezels on the laser cutter (Figure **F**).

FEATHER FONA PINS: 3 5 6   +V   GND

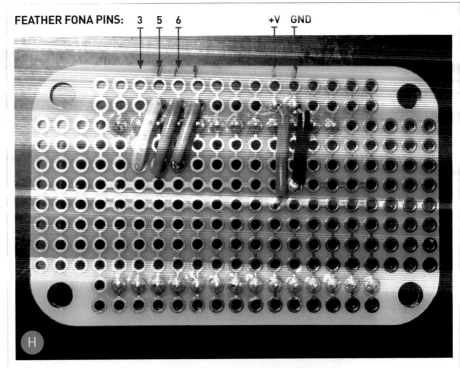

Mask the threaded inserts with tape (Figure **G**) before placing them into the case, then epoxy them into their holes in the top part of the case. Careful, you don't want the epoxy to seep into the inside threads.

Super-glue the light panel to the back of the yellow speaker bezel, then glue it and the microphone bezel in place.

Glue the optional back bezel "plaque" in place with super glue. I created and printed a Motorola logo badge for this — use an antique font to generate your own.

ASSEMBLE THE ELECTRONICS
Prepare the circuit board. You'll have to cut some traces and add jumpers due to the required positioning of the OLED display and the Feather Fona (Figure **H**).

CRITICAL: You need to desolder the straight header pins on the keypad and solder in right-angle headers instead. Otherwise, things won't fit.

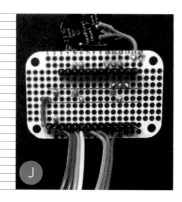

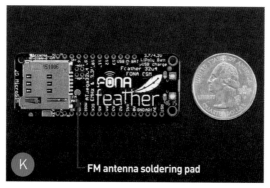

FM antenna soldering pad

Solder mic here →

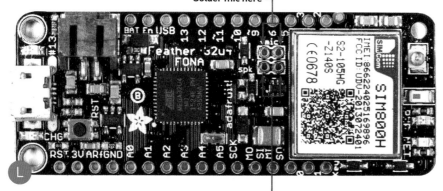

Solder speaker here →

Feather Fona
USB

Feather Fona
Pin 2

Align the OLED display into its cutout and hot-glue it down, then do the same with the keypad. I friction-fit some "vintage" grille fabric in place behind the grill bezel (Figure **I**, previous page).

Solder jumper wires to the Perma-Proto board for the keypad. Also solder the NeoPixel Jewel to the board (Figure **J**).

If you want to use the FM radio, solder a 3.1-foot length of 22AWG wire to the back antenna pad (Figure **K**). You'll wrap the antenna around the keypad and other components in the top of the case. Because of the tiny speaker's range (~600Hz-10kHz), speech sounds fine, but FM radio sounds bad, missing the highs.

Optionally, you can solder Adafruit LED sequins to a chunk of Perma-Proto board (just +V/GND) to backlight the microphone bezel. If you go that route, remember to solder them before mounting the board in place.

The Feather Fona board has connection points for a microphone and speaker (Figure **L**) — solder yours to these by feeding the wires in through the top, adding solder on the bottom of the Fona.

Connect the JST extension to the battery jack. Thread the battery/JST extension under where the speaker will sit. This extension connects to an on-off switch on the back case.

Test-fit the Feather Fona onto the OLED pins. The USB pin should be at pin 5 of the Perma-Proto (Figure **M**).

Solder the Feather Fona in place. Cover soldered pins under the speaker with electrical tape to prevent shorts. Fit the speaker down into the speaker cutout and hot-glue it in place.

Cut the JST extension to attach to the 12mm on-off switch on the back. Solder it to the switch and connect the battery (Figure **N**). Cover the soldered pins with electrical tape to prevent shorts, then test the switch to make sure it works. Super-glue the switch in place. Final assembly should look something like Figure **O**.

MAKING CALLS

To use the cellphone, push the red on-off button on the back. The microphone will light up and you'll see a brief "Welcome to RadioPhone" splash screen. Next, the screen displays "Looking for Network." If the phone connects to the carrier, you'll

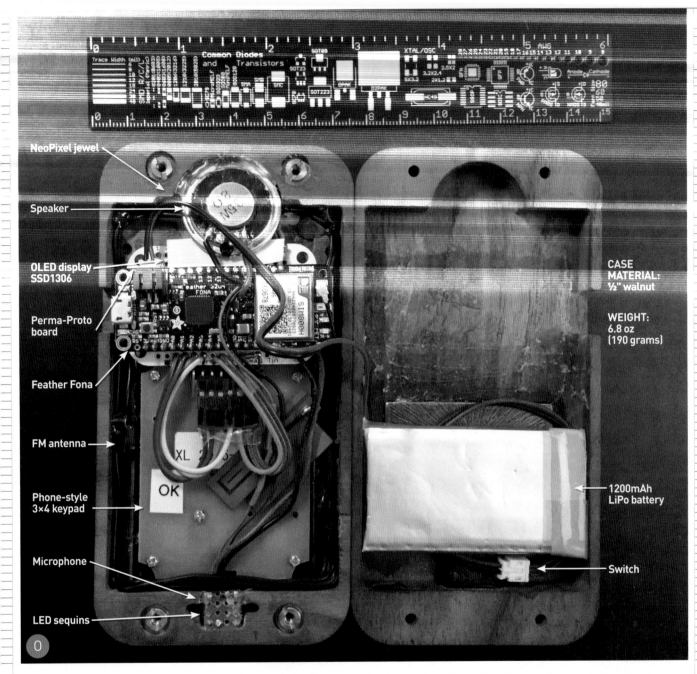

NeoPixel jewel

Speaker

OLED display
SSD1306

Perma-Proto
board

Feather Fona

FM antenna

Phone-style
3×4 keypad

Microphone

LED sequins

CASE
MATERIAL:
½" walnut

WEIGHT:
6.8 oz
(190 grams)

1200mAh
LiPo battery

Switch

see the message: "Connected to Network!" for a few seconds. The speaker backlight will illuminate and the menu will appear. Otherwise, with no carrier signal, you'll just be stuck on "Looking for Network."

The menu screen displays signal strength in the upper left corner and battery level/% charged in the upper right corner. The date and time from the carrier network shows up in the center top of the screen. Below that are the function choices. You can dial a number (press 1 on the keypad), call one of five favorite numbers (press 2), or listen to one of five pre-stored FM radio stations (press 3). In the lower

right corner, there's a # (hash/pound sign). An incoming call will be signaled by tones; just press # on the keypad from the menu screen to answer it. There's also a * (star/asterisk) in the lower left corner. Toggle this for a flashlight — it turns on all seven LEDs under the speaker grille.

To make a call, press 1, enter a number on the next "Number Please!" display, and press *. You can back out of this screen by pressing #. To hang up for outgoing or incoming calls, press #. If you want to call one of your favorites, press 2, then the number of the favorite to call from the next displayed screen. You have to pre-store

these by changing code in the sketch. Pressing 3 brings up your favorite FM radio choices, also stored in code (no decimals included — "98.1 FM" becomes "981"). Press one of the numbers shown to play that radio station.

When you're done using your phone, just push the red button on the back again. Instant off, no power down sequence. Gratifying! ⊘

For detailed build steps, videos, and more, visit thisoldgeek.blogspot.com.

Teeny-Tiny
Spy Bug

Written by Tom Schneider

Build the world's smallest FM transmitter for all your espionage needs

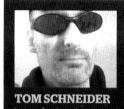

TOM SCHNEIDER is an electronics technician from Canada. He's interested in all kinds of new technologies and loves to spend his time with unusual projects.

Time Required: About 3 Hours
Cost: $10–$20

MATERIALS

» **Printed circuit board** Download the Gerber files from the project page at makezine.com/go/tiny-fm-spy-transmitter, and order your boards from a PCB vendor; I used OSH Park (oshpark.com).

» **Surface-mount components** See the Required Components table at right.

TOOLS

» **Soldering iron with fine tip** smaller than 0.8mm

» **Microscope** with at least 20x magnification. An inexpensive USB microscope is sufficient.

» **Solder paste**

» **Fine solder wire** ideally smaller than 0.5mm

» **Fine tweezers**

» **Bench belt sander, flat file, or Dremel** for removing excess material from PCB

» **Steady hands!**

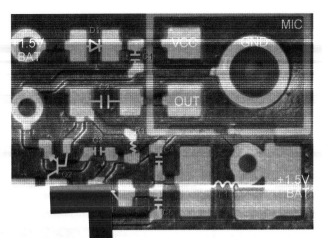

Schematic labels:

Max 1.55V VCC

D1 BAT54LPS-7

C1 100nF

MIC ICS-40310 OUT

C2 2.2µF

C3 100nF

R1 1M

C4 12pF

L1 220nH

C5 5.6pF

Q1 BFR360L3

TX

Q2 MMBT3904SL

A

VDD

GND

WANT TO EAVESDROP...
AGENT? HERE'S HOW

SPY BUG with extremel... that transmits an FM si... a voltage of only 1.5V. Th... that's proven to work rel... described here, but it's a... — literally microscopic! Y...

NOTE: Once your bug is b... responsible for observing l... regulations for operating s...

THE CIRCUIT

The microphone on the le... schematic diagram (Figur... maximum voltage rating o... battery voltage must there... down to a lower value, sinc... cells have a higher nomina... 1.45V. That's the purpose of... diode D1, which at the extr... microphone current of appr...

...y, but this makes no sense because ...est currently commercialized ...ll has a diameter of 4.8mm. ...t all of the required components ... below left) from Digi-Key (digikey. ...you can choose other suppliers ...ouser. There are no exact model ...or the resistors and capacitors ...ese differ by supplier, and ...e over time. For the passive ...ts only the capacitance/resistance ...rse the overall size are important. ...-determining capacitors should ...ature-stable types such as NPO. ...ncy-determining components are ...e for a frequency of 81MHz. For ...ays select a component with an ...d a high Q factor (Figure **B**).

...wnload the Gerber files from ...page at makezine.com/go/tiny-...smitter, then send them to any

...ATIONS OF THE SPY BUG

...ns: 4.75mm×6.8mm ...268")

...ply: Minimum 1.1V, maximum ...mally supplied by a silver-oxide

...nsumption: <200 micro amps

...time: Depends on battery ...t 48 hours using the ...ommercially available battery

»Range: For optimum range, an antenna ¼ wavelength long (e.g., 88cm at 85MHz) transmits about 50m in open terrain. Shorter antennas are possible with certain losses.

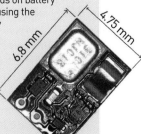

4.75 mm
6.8 mm

Required Comp...

COMPONENT	MODE...		
MIC	MEMS a...		
D1	Schottk...		
C1, C3	Ceramic...		
C2	Ceramic...		
R1	Resistor,...		
C4	Ceramic capacitor, 12pF, NPO	01005, 0.4mm×0.2mm	
C5	Ceramic capacitor, 5.6pF, NPO	01005, 0.4mm×0.2mm	
L1	Inductor coil, 220nH, wire wound, air core (LQW2BASR22J00L)	0805, 2.09mm×1.53mm	
Q1	Transistor, NPN (BFR360L3)	TSLP-3, 0.6mm×1.0mm	
Q2	Transistor, NPN (MMBT3904SLCT)	SOT923F, 0.6mm×1.0mm	
Battery	1.2V-1.55V	SR416SW, 1.6mm×4.8mm	

Tom Schneider

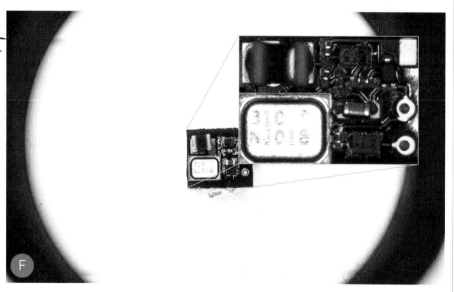

ASSEMBLY

To assemble the components on the circuit board, you'll need to use a microscope. I recommend you watch my assembly video at youtube.com/watch?v=_Na0Ac7FTbk before you begin. The board is very lightweight, so you should fix it to a surface before doing the following steps.

First, deposit a little solder paste on each pad using a needle or a piece of thin wire (Figure E). If you accidentally put too much on the pad, remove the excess before soldering. Then carefully align and place the components on the pads (Figure F). Push them down slightly to prevent them from slipping off.

Now heat the board to about 480°F (250°C) on a hot plate or in a small oven. If you use an oven, it should be one without strong air movement, otherwise the parts may simply be blown off the PCB.

When the correct temperature has almost been reached, smoke will emanate from the solder paste. The beads of solder will then melt and connect the components to the printed conductors (Figure G). Once this happens, turn off the heat (and open the oven door).

Allow the board to cool down a bit before handling. If you're using a hot plate, immediately move the board off the hottest spot in order to avoid damage to the MEMS microphone, which is pretty sensitive.

If you're using the SR416SW button battery, you can attach a battery clip, or solder a short wire into the 2 through-holes (vias) and use it as a DIY battery clip to connect the positive battery terminal to the circuit board (Figure H). Of course you can also solder wires to the GND and +1.5V pads to connect another type of battery.

OSCILLATION FREQUENCY AND ANTENNA

I chose the 81MHz frequency deliberately because in many countries the FM range between 87.5MHz and 108MHz is overcrowded, which makes it more difficult to receive the relatively weak signal of our transmitter. Use a scanner or a world radio to hear your spy bug.

In some countries radio stations broadcast at 76MHz–88MHz, and smartphones containing FM receivers can be adjusted to use this frequency band.

PCB manufacturer. For such a small circuit board, I recommend using a company like OSH Park that puts together orders from different customers on a single large board that is then sent to a PCB manufacturer. The cost of each small PCB goes down considerably.

Figures C and D show the board as it comes from the manufacturer. They have

been panelized and then broken out, so you'll need to remove some excess material around the edges using a bench belt sander (a flat file or Dremel will work too). This circuit board probably can't be produced at home using toner transfer or similar methods because the traces are only 0.2mm wide and the distance between them is equally small.

Ideally the antenna length should be at least ¼ of the wavelength. Here's an example of a configuration the circuit has been tested with:

C4	C5	L1	Frequency
12pF	5.6pF	220nH	81MHz

Antenna length	Range (open field)
92cm (36")	160ft

If a different oscillation frequency is chosen the antenna length should be adjusted accordingly. The wavelength can be calculated according to the following formula:

$$\lambda = v / f$$

where λ is the wavelength, v is the speed of light, or about 300,000km/s, and f is the frequency.

The antenna (Figure I) can be made shorter than ¼ wavelength if you're willing to accept a shorter range. To use a short "whip" antenna, connect an inductance of a few tens of nH between the antenna wire and the transmitter. The value of this inductor is best determined experimentally.

MODIFICATIONS
» Instead of the MEMS microphone, you can use an electret condenser microphone, and omit the diode.
» If the incoming audio signal is too strong,

place an additional resistor in series with C2. The value of this resistor needs to be determined experimentally.
» If low-frequency noise, like from a fan, is present in the room under surveillance, the value of C2 can be decreased. The circuit works very well with values down to 100nF.

I'm going to keep improving this project, so please check back on the project page online. Meanwhile, if you have questions, email me at tomtechtod@gmail.com. And now, happy experimenting!

Check out Tom's latest project: A 915MHz UHF SMD spy bug that fits in a common pen at youtube.com/watch?v=wMkaN21K5S0.

Share your spy bug build and see more photos at makezine.com/go/tiny-fm-spy-transmitter.

Come Glide With Me

Mod your Air Rocket Launcher to make your aircraft soar

Written by Katherine Ozawa and Rick Schertle

RICK SCHERTLE runs the maker lab at Steindorf K-8 STEAM School in San Jose, California. He's a contributing writer to Make:, author of *Make: Planes, Gliders and Paper Rockets*, and co-founder of Air Rocket Works. Along with his wife and kids, he loves all things that fly.

KATHERINE OZAWA graduated from Swarthmore College with a B.S. in engineering before starting her career as an educator. She now combines her backgrounds to create museum experiences focused on engaging kids with engineering problems in tangible and exciting ways.

Time Required:
1-2 Hours
Cost:
$109 Launcher;
$3–$5 Glider
(depending on materials used)

MATERIALS

GLIDER:
- » **ABS plastic tube, 10"** from Air Rocket Works, airrocket works.com, for glider fuselage
- » **Squishy foam nose cone** from airrocketworks.com, for fuselage nose
- » **Wing material** a variety of materials can be used
- » **Rigid foam insulation** We use Owens Corning Foamular Fanfold Insulation Sheathing.
- » **Disposable foam plates**
- » **Cardboard**
- » **Foamcore**
- » **Plastic sheeting** Kitronik kitronik.co.uk
- » **Velcro, ¾"** with adhesive backing (optional)
- » **Rubber bands**
- » **Washers, ¼"** for weight
- » **Clear tape**

LAUNCHER:
- » **Compressed Air Rocket Launcher Kit** includes ½" PVC launch tube, airrocketworks. com or makershed.com
- » **Steel NPT street elbow, ½"** Available from any hardware store, airrocketworks.com, or McMaster-Carr #44605k134, mcmaster.com
- » **Pipe sealing tape** comes with launcher kit

TOOLS

GLIDER:
- » **Glue gun**
- » **Scissors**

LAUNCHER:
- » **Adjustable wrench**

Hep Svadja

THE TECH STUDIO IN THE TECH MUSEUM OF INNOVATION IS A FLEXIBLE INTERACTIVE SPACE intended to inspire the maker in everyone through design challenge learning. In creating Just Wing It, a recent studio project, I wanted to draw out the wonder that flight inspires in all of us.

In Just Wing It, guests could choose from a variety of components to design unique flying contraptions. In order to encourage creative design, the exhibit contained a wide library of biomimetic and whimsical wings. These wings are added to fuselages that were built off of Rick's nylon Air Rocket Gliders.

For the testing rig, I took the Compressed Air Rocket Launcher and expanded it to create a two-person rig where guests could see their creations fly simultaneously. The modular glider design ensured that guests could quickly iterate after each launch to test multiple design ideas during the experience.

The best part of Just Wing It was seeing guests go from a less than successful design to gliders that reached the ceiling. It wasn't uncommon to hear cheering from that section of the museum floor!

—*Katherine Ozawa*

Since I first introduced the Compressed Air Rocket Launcher to the world in *Make:* Vol. 15 in 2008 it's been incredibly popular. After a while, I realized it could be used not just as a rocket launcher, but a launch rig for other things as well.

I finally had a chance to try this out when approached by The Tech Museum to help with an exhibit last year. This glider launch rig (Figure **A**) utilizes our latest version of the Compressed Air Rocket Launcher, a solid wood-and-metal industrial design. I give a huge amount of credit to my partner at AirRocketWorks.com, Keith Violette, for his design and manufacturing brilliance. Since designing the exhibit for The Tech Museum, we now have a custom ABS tube and squishy foam nose cone that fit on the standard ½" PVC launch tube, making this project even more accessible to individuals, teachers, and Maker Camp directors.

Follow these simple steps to create a whole new design and prototyping experience using the Compressed Air Rocket Launcher.

—*Rick Schertle*

BUILD YOUR GLIDER — QUICK BUILD WITH VELCRO

Several types can be built. The velcro model was used at The Tech Museum for quick iterations for museum guests.

1. Insert and tape the squishy nose cone to the precut lightweight ABS tube (Figure **B**). If you don't tape it on securely, it will blow off when your glider is launched, even at 10psi or 15psi. If you forget to tape it on, you can also glue it on with super glue or a hot glue gun.

2. Cut four 10" lengths of adhesive velcro tape and stick to the fuselage (Figure **C**).

3. Attach the other half of the velcro tape onto your wings of choice. Figure **D** shows the durable plastic wings used for thousands of launches at The Tech Museum exhibit. Now kids can quickly swap a variety of wings on and off for rapid experimentation.

BUILD YOUR GLIDER — MIXED MATERIALS

1. Insert and tape the squishy nose cone to the precut lightweight ABS tube, as in the previous project (Figure B).

2. Build your wings out of a variety of materials. You can use the template or laser cut files found in this project online (makezine.com/go/glider-launch-rig), or design one with your own shape.

3. To make the main wing, you can hot-glue the two pieces together with a dihedral angle as shown in Figure **E**.

4. Glue the rudder to the elevator (Figure **F**).

5. Rubber-band the wings to the fuselage as shown in Figure **G**. With the force at launch, you might find the fuselage firing off and leaving the wings behind. If so, you might need to make some tweaks.

6. Your finished glider (Figure **H**, following page) is ready for launch!

This project has tons of flexibility in both materials and scale. If you're building with younger kids, the velcro works great because kids can add and swap out various

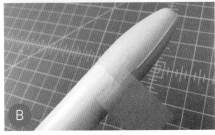

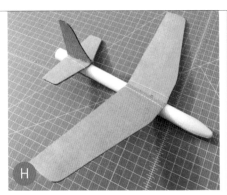

wings and materials. Simply pre-make lots of fuselages and wing materials and then let the kids go to town! This is how it ran at The Tech Museum with hundreds of gliders being built and flown each day.

BUILD THE GLIDER LAUNCHER

1. Build the Compressed Air Rocket Launcher but don't add the ½" launch tube (Figure **I**).

2. Add pipe seal (teflon) tape to the threaded male side of the NPT street elbow (Figure **J**), then thread it into the top of the QEV valve — the port labeled "R". The street elbow is necessary because the QEV needs to operate at a vertical angle so the rubber diaphragm inside resets after each launch.

3. Tighten the street elbow into the QEV valve so it points away from the launcher.

4. Now screw the ½" gray PVC launch tube to the street elbow so that it sticks out at a 90° angle (Figure **K**). The connection between the street elbow and PVC tube does not need any pipe sealing tape. You can adjust the angle of the launch tube by loosening and then tightening the wing nut on the launcher that holds the aluminum pressure chamber in place.

5. Attach a bicycle pump or compressor with a pressure gauge to the launcher. Slide your glider prototype onto the launch tube (Figure **L**). Set the slide valve to the "Pressurize" position and add 10psi–15psi of pressure. This should be *plenty* for launches of 50'–100'. Move the slide valve back to the "Launch" position and watch your glider shoot off!

This is a prototype rig, so adjust your glider as you go. Experiment with different wing sizes and materials, using the ABS tube and foam nose as your building platform. ◗

Adjusting your gliders to fly better

Your glider should fly level with a gentle climb. If it stalls or dives, try the following:

» If your glider noses up, stalls, and then falls to the ground, you can try moving the main wings forward or adding weight to the nose. A neat and clean way to add weight is to take off the nose cone and tape one or more ¼" washers inside the nose cone before putting it back on again.

» If your glider goes right into a dive, you need to move the main wings back and/or add weight to the back of the glider.

» If your glider flies level, but banks to the right or left, try adding a moveable cardboard (or post-it note) rudder or ailerons.

Your glider is all about trial and error and learning what it takes to get it to fly straight and level.

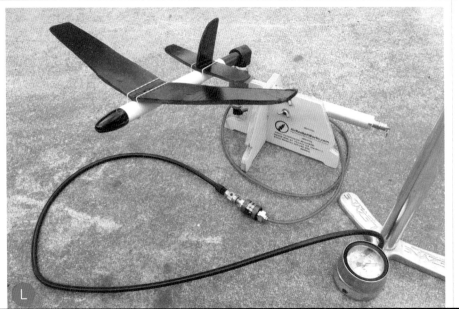

Get the template or laser-cut files and share your glider launcher builds at makezine.com/go/glider-launch-rig.

Rick Schertle

Just in Case

Create a custom foam insert to transport your fragile equipment

Written by Charles Platt

CHARLES PLATT
is the author of
Make: Electronics, an
introductory guide
for all ages, its sequel
Make: More Electronics,
and the 3-volume
*Encyclopedia of
Electronic Components*.
His new book, *Make:
Tools*, is available now.
makershed.com/platt

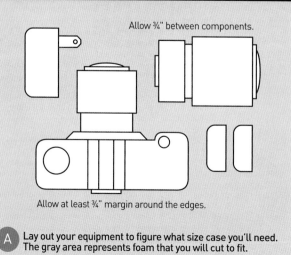

Allow ¾" between components.

Allow at least ¾" margin around the edges.

A Lay out your equipment to figure what size case you'll need. The gray area represents foam that you will cut to fit.

B Some examples of Pelican brand transport cases.

Charles Platt

TRANSPORTING DELICATE ELECTRONIC OR PHOTOGRAPHIC EQUIPMENT IS EASY TO DO. Just embed it in squishy polyurethane foam inside a rigid case.

The problem is achieving this without spending too much money.

The first step is to figure out how big a case you need. Lay out your equipment as suggested in Figure Ⓐ, allowing at least ¾" between each piece and around the edges. Make a note of the minimum area that will hold everything. Next measure the tallest piece of equipment and add at least 1½". This tells you the minimum internal height of the case.

Now that you have your measurements, you can check an online vendor to find a case that's the right size or slightly larger. Pelican brand cases are my favorite. They are almost indestructible and are available in a great variety of sizes and colors, such as the samples in Figure Ⓑ. The chart at pelican-case.com/chart.html shows all the exterior and interior dimensions. Make a note of model numbers that are suitable for your purposes.

New Pelicans are expensive, but because they are so robust, you can feel fairly confident about buying them secondhand. Search eBay for the model numbers that you want, and don't be put off if a case looks grungy. Used cases often have an accumulation of stickers on the outside, but you can remove them with a paint scraper while applying a heat gun or hair dryer. Any remaining residue should yield to a solvent such as Xylol, but wear latex or nitrile gloves, work with good ventilation, and don't leave the solvent on the case for too long. If it dulls the finish, you can restore the appearance by scrubbing it with a fine wire brush and then polishing it.

If there are dividers inside a used case, you can discard them. I think it's better to use polyurethane foam that fits your equipment precisely. Foam is available for this purpose in "pick-and-pluck" form, meaning that it is partially cut into small square-shaped pieces that you pull out to make room for your gear. Personally, however, I prefer to buy a large generic piece of foam and custom-cut it myself. A fabric store such as JoAnn sells suitable foam that is intended for chair cushions.

You'll need three pieces: two relatively thin slabs to go inside the lid and the base, and a thicker block to go between them. Figure Ⓒ shows a typical end view. To figure the thickness of the top and bottom pieces, take the internal height of the case, subtract the height of your largest piece of equipment, divide by 2, then add ½" to each piece so that they'll grip the equipment when the case is closed.

Cutting foam is more difficult than you might think. A wood saw will make a mess. A hacksaw is better, but hard to control. A kitchen knife tends to get stuck. A bread knife with fine serrations works quite well, but you'll still face a challenge if you want a cut that is precisely vertical. An electric carving knife is the ideal tool for long, straight, deep cuts. Clamp the foam gently between a piece of plywood and your workbench, leaving a section sticking out that you want to remove. Wearing work gloves is a sensible precaution, and, of course, never turn any kind of blade toward you.

After cutting the middle piece of foam, you need to make holes in it for your equipment. Draw around each item with a marker, then cut ¼" inside the line, so that items will fit securely but can still be inserted and removed without too much effort. A hand-held coping saw is a cheap and simple way to cut curves, and my new book, *Make: Tools* will tell you how to use it, along with many other hand tools.

Figure Ⓓ shows cutouts for a camera and battery charger. Figure Ⓔ shows foam installed in a small case, and the main image on the facing page shows the case packed and ready to go.

If cutting foam to fit seems challenging, there's a simpler option. You can cut simple blocks of foam and push-fit them together, as shown in Figures Ⓕ and Ⓖ. The advantage of this system is that you can shuffle the blocks around if you buy new equipment of different sizes.

Whichever system you use, you'll have a feeling of freedom when your expensive gear is properly protected. You can dump the transport case into the back of a vehicle without worrying if it slides on the floor or gets buried under other items. You may even feel sufficiently confident to send it as checked baggage on an airline. Remember that although Pelican cases allow for use of padlocks, TSA regulations require you to use a special approved lock to allow for inspection en route. ⊘

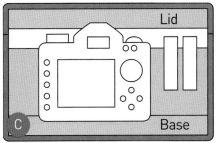

End view of transport case showing a camera and two batteries embedded in foam.

A coping saw can cut holes for equipment such as a camera and battery charger.

The space on the right-hand side allows room for accessories such as wires that don't require so much protection.

Cutting movable pieces of foam is a simple option.

If you buy different equipment, you can rearrange the foam pieces.

Better Times

Written by Larry Cotton

How I made an accurate, easy-to-read clock using stepper motors

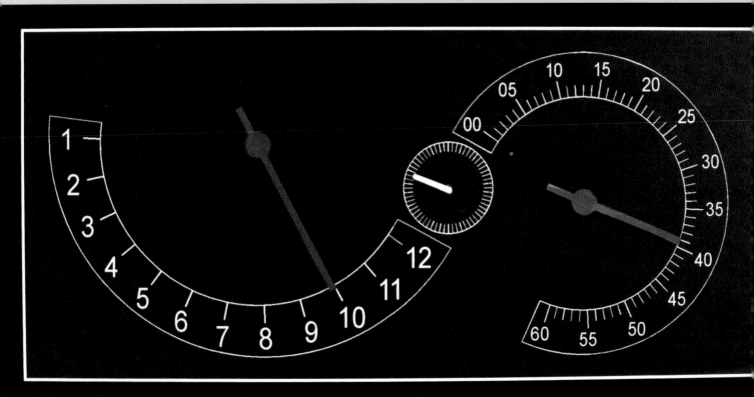

ANCIENT CLOCKS WERE MECHANICAL, AND EVOLVED WITH ROUND FACES PRIMARILY FOR THAT REASON
(Figure **A**). When electricity became widely available, clocks' faces copied (and still copy) the old mechanical ones because it was easy and cheap to rotate the hands. So we've got used to it, and today most clocks still have a round face with two or three hands that rotate. OK, there's those boring digital clocks . . .

Then along came stepper motors whose shafts still rotate, but in tiny steps, in either direction, at easily controllable speeds. So why don't clocks use them? They would not need circular dials but would be just as accurate. It's much easier to tell time on a stepper clock, too. (How long do kids spend learning how to "tell time"?) And they're very cheap: how about 6 bucks for a set of 5 stepper motors and control boards

from Amazon?

Check out Figure **B**. What time is it? This clock is powered by two stepper motors which move the hands along easy-to-read arc-shaped minute and hour dials. I considered moving at least one hand in a straight line (with a timing belt or rack and pinion — Figures **C** through **E**) but after much consternation and cogitation I settled on two arcs in an S-shape. I added a cheap clock mechanism ($.50 at AliExpress) with a lone second hand, mainly to show that the clock is running.

THE ELECTRONICS

My clock is timed by another cheap clock. Its electronics include an IC fed by a quartz crystal running at 32,768 cycles per second. A binary digital counter driven at that frequency overflows once per second, creating alternating positive and negative pulses to drive its solenoid. A positive pulse every two seconds is fine for driving stepper motors via proper programming and electronics. One caveat: the cheap clock must tick, not hum. Check out en.wikipedia.org/wiki/Quartz_clock and explainthatstuff.com/quartzclockwatch.html

Here's a YouTube video (youtube.com/watch?v=XzXfadQXRn8) that shows (about 3 minutes in) how to get that positive signal from a cheap clock. Connecting signal and ground wires to the two solenoid connection points should yield 1 positive pulse every two seconds which can be sent to a programmable circuit board such as Arduino, Basic Stamp, etc.

An alternative would be to use the circuit board's internal clock, but I discovered programming is easier, and a stepper clock is potentially more accurate, if timing comes from one of the (ticking) cheap clocks.

I'm a relative newbie to steppers, so I read tons of information and watched zillions of YouTube videos to try to divine their workings. Basically, they're multi-coil motors whose shafts turn — in either direction — in tiny steps. Each step requires an electrical pulse to turn the shaft just a little bit. Here's a good tutorial and information on the specific motor I used: learn.adafruit.com/all-about-stepper-motors and adafruit.com/product/858. The company offers lots of stepper-motor related products and help.

My clock uses two 28BYJ48 steppers to turn its minute and hour hands. The 28BYJ48 produces 32 steps per full *motor* revolution. Its output shaft is geared to yield 512 (or 513, depending on source) steps per revolution. For those familiar with Arduino programming and shields, you can find help programming stepper

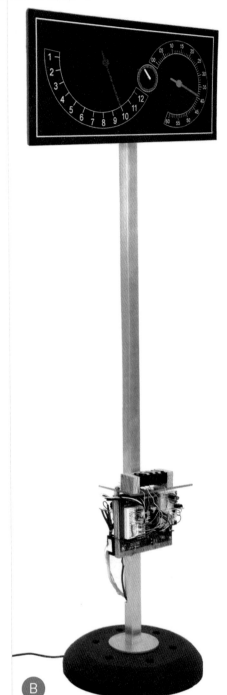

(B)

LARRY COTTON is a semi-retired power-tool designer and part-time community college math instructor. He loves music and musical instruments, computers, birds, electronics, furniture design, and his wife — not necessarily in that order.

Time Required: A Weekend
Cost: $100–$150

MATERIALS

- » **Stepper motors, small (2)** such as Amazon #B00EYVH6GC amazon.com
- » **Stepper motor driver board, ULN2003 (2)**
- » **Programmable circuit board** such as Arduino, Basic Stamp, etc.
- » **Power supply, wall wart, rated at 9-12VDC and 1 Amp (1000 mA) minimum**
- » **Prototype breadboards, small (several)**
- » **Standard electronic components**
- » **22-gauge connecting wire, plugs, and sockets (lots)**
- » **Switches, SPST**
- » **Cheap clock mechanisms that tick (several)** check out AliExpress.com
- » **Solder**
- » **Plywood ¼"**
- » **Cheap floor lamp** for a stand if you're building a floor stand clock
- » **Quality print of your clock face**
- » **Thin scissor-cuttable plastic** for hands
- » **Paint** for hands and back
- » **Spray adhesive,** such as Loctite Repositionable Spray Adhesive
- » **Miscellaneous fasteners**
- » **Adhesives — super and hot glue, masking and double-sided tape, etc.**

TOOLS

I have many tools and materials on hand, which greatly sped up the construction of my stepper clock. Some of my more useful are:

- » **Shopsmith** (which I use mostly as a disc or drum sander)
- » **Band saw** (also Shopsmith but on a separate motor and stand)
- » **Small bench drill press with 0"–½" chuck**
- » **Portable drill/screwdriver, cordless, 0"–⅜" dia. chuck capacity, variable speed, reversible**
- » **Drill bits 1⁄16" through ½"**
- » **Jigsaw, hacksaw, and blades**
- » **Scissors**
- » **Wire strippers**
- » **Small soldering iron**
- » **Common hand tools** such as a hammer, lots of pliers, and screwdrivers
- » **Sandpaper**

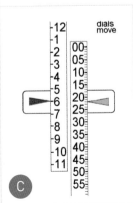

(C)

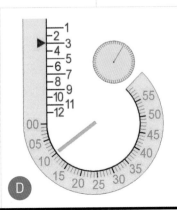

(D)

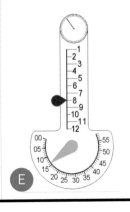

(E)

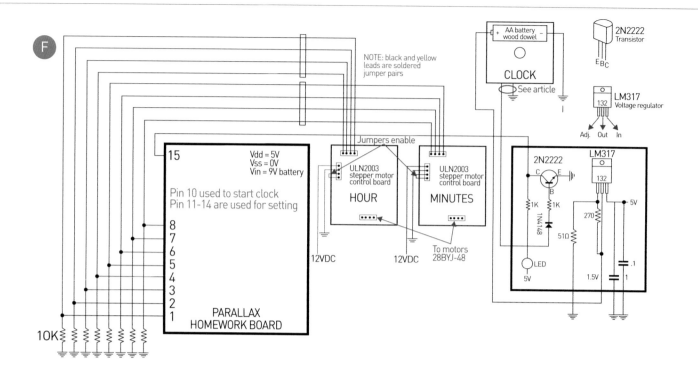

F

NOTE: black and yellow leads are soldered jumper pairs

AA battery wood dowel

CLOCK

See article

2N2222 Transistor
E B C

LM317 Voltage regulator
Adj. Out In

15

Vdd = 5V
Vss = 0V
Vin = 9V battery

Pin 10 used to start clock
Pin 11-14 are used for setting

Jumpers enable

ULN2003 stepper motor control board
HOUR

ULN2003 stepper motor control board
MINUTES

2N2222

LM317

8
7
6
5
4
3
2
1

PARALLAX
HOMEWORK BOARD

12VDC

12VDC

To motors
28BYJ-48

10K

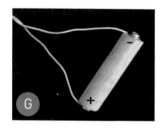

G

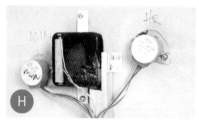

H

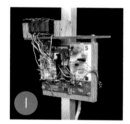

I

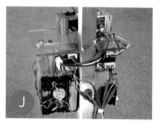

J

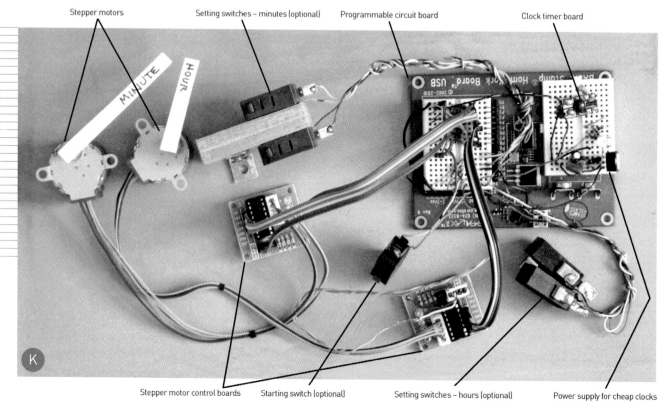

Stepper motors

Setting switches – minutes (optional)

Programmable circuit board

Clock timer board

MINUTE

HOUR

K

Stepper motor control boards

Starting switch (optional)

Setting switches – hours (optional)

Power supply for cheap clocks

Larry Cotton

motors here: arduino.cc/en/Reference/Stepper?from=Tutorial.Stepper.

No matter the source of the timing pulses, the board should be programmed to wait for 30 of them, then quickly move the minute-hand's stepper motor from one minute to the next. After 60 minutes, a signal must go to the hour hand's stepper to move it from one hour to the next. (The hands on my clock are attached directly to the stepper shafts.)

It's visually satisfying to see the hands move quickly between minutes and hours: for example, at exactly 8:00 the hour hand will move from 7 to 8. When the hands get to the ends of their respective dials, they quickly reverse to their beginning points.

I programmed my steppers for 6 quick steps from minute to minute and 20 steps from hour to hour. Your mileage will vary depending on your clock-face graphics. If you want to use my clock face, you can find it in this project online at makezine.com/go/stepper-motor-clock.

See Figure F for my circuit. It shows a Parallax HomeWork Board, but connections to an Arduino would be similar. I included a power-supply to send 1.5VDC to both cheap clocks; ½" wood dowels with screws in the ends replace their AA batteries (detail in Figure G). Any deviation from 1.5VDC will necessitate a change to the 51-ohm resistor.

THE SOFTWARE

Whether using C++ or PBASIC, experiment with values for pause t to get a smooth but fast hand move. Too short pauses may cause the hands to stutter. Too long pauses take too much time to move the hands.

This subroutine (in PBASIC) quickly steps the minute hand clockwise, using HW Board's input pins 1-4:

```
FOR z=1 TO 6
HIGH 1: LOW 2: LOW 3: LOW 4
PAUSE t
LOW 1: HIGH 2: LOW 3: LOW 4
PAUSE t
LOW 1: LOW 2: HIGH 3: LOW 4
PAUSE t
LOW 1: LOW 2: LOW 3: HIGH 4
PAUSE t
NEXT
LOW 1: LOW 2: LOW 3: LOW 4
```

The first line begins a loop to step the minute hand enough times to move from minute to minute on the dial. The last line deactivates the stepper motor coils to keep the motor cooler and to draw less current.

Since I chose to move the hands in opposite directions, stepping the hour hand required reversing the previous coil activation sequence using pins 5–8; use enough steps to move the hand from hour to hour.

Returning the hands to 00 and 1 requires sending the *cumulative* number of forward pulses — in reverse order — to the steppers. In my case I used 360 (60 minutes x 6 steps per minute-hand move) and 240 (12 hours x 20 steps per hour-hand move) reverse steps.

THE MECHANICALS

I created my clock face in AutoSketch, including tiny circles to locate holes for the stepper and cheap-clock shafts; then I saved it as a PDF file. I had the clock face printed at Staples, because I didn't want to drain an entire black-ink printer cartridge. Bonus: this face (13"×6½") cost only 20 cents to print, so I added four backup prints!

Next came light sanding and spray-painting both sides of a piece of good quality ¼" plywood, spray-gluing the face, trimming the edges, and drilling shaft holes. I screwed the stepper motors to the back of the plywood with the shafts poking through their holes and mounted the other seconds-only cheap clock (Figure H).

I chose to build a floor clock and found a lamp at Target from which I cannibalized the stand. Since my electronics aren't exactly, um, professional quality (Figures I and J), I mounted them low on the stand and ran the wires through the stand tubing.

If the electronics were arranged more compactly, they could be mounted on the back of the plywood face for a desk or wall clock configuration. I've shown all the parts in Figure K.

Lazy builder/programmer tip: set the clock by turning it off, positioning the hands at 1 and 00, and restarting it at exactly 1 PM. Refer to Tip 3. Night owls may disagree. ✪

TIPS

1. It's best to create your clock's face based on actual movement of the hands. Program the minute hand to cover its arc in an hour, and the hour hand to cover its arc in 12 hours. To speed up the process, give each minute only two seconds!

2. Be sure to turn your steppers off (all low signals) when not moving hands. The motors will run cooler and draw much less current.

3. Attach the hands with a short piece of flexible tubing that's press-fit over the stepper shaft. If you mount them rigidly, extra coding at the beginning of the program will be required. In addition, setting and starting switches will be necessary (and are in fact shown in Figures I-K).

4. Ensure your power supply supplies adequate current to the steppers. The small steppers I used don't draw much current, but bigger ones would, especially if they have to hold a position against a load.

5. Use plugs and sockets on long wires through floor stand tubing. I regret not doing this!

Get clock face graphics and share your stepper motor clock builds at makezine.com/go/stepper-motor-clock.

Written by Paloma Fautley

Culinary Chemistry

Make your bruschetta better with cold-oil spherification

MOLECULAR GASTRONOMY — ALSO KNOWN AS MULTI-SENSORY COOKING, modernist cuisine, culinary physics, and experimental cuisine — has led to many innovative dining experiences.

This study of the chemical transformation of the tastes and textures of food through cooking has a chef-centric approach instead of a science-centric approach, which makes it more useful to the everyday cook.

Understanding even the basic interactions between common ingredients can lead to more consistent and successful cooking.

One recently popular process is spherification, which creates small gel spheres resembling caviar that pop with flavor when eaten.

Traditional spherification uses sodium alginate and calcium to create delicate molecular droplets, but in this project we will employ a much simpler technique using agar and cooled oil. You can imagine agar like gelatin, which is liquid at high temperatures but solidifies when cooled. We take advantage of this property by dripping the hot mixture into cold oil so that it forms a gelatinous sphere as it rapidly cools. These "caviar" can then be added in place of a traditional splash of balsamic for a fun textural treat!

Hep Svadja

PALOMA FAUTLEY is the co-chair of her IEEE student branch and is pursuing a degree in robotics engineering at the University of California, Santa Cruz.

Time Required: 1 Hour
Cost: $15–$25

INGREDIENTS

FOR THE BALSAMIC CAVIAR:
» **100g balsamic vinegar**
» **1.5g agar**
» **1 cup olive oil, chilled**

FOR THE BRUSCHETTA:
» **Baguette**
» **2 medium tomatoes, diced**
  I prefer heirloom
» **3 large garlic cloves, minced**
» **1 small red onion, diced**
» **The juice of 1 lemon**
» **1 cup fresh mozzarella, sliced**
» **½ cup whole basil leaves**
» **Olive oil**
» **Salt and pepper to taste**

TOOLS
» **Syringe**
» **Whisk**
» **Instant read thermometer**
» **Kitchen scale, accurate to at least 0.1gram**

More Molecular Gastronomy

Cream whippers can be exploited for many other culinary creations, such as rapid flavor infusions and marinades, in a method called nitrogen cavitation.

HOW IT WORKS
Nitrogen cavitation homogenizes cells and tissues using rapid decompression of gases. Add nitrous oxide gas (N_2O) to your ingredients in the cream whipper and release the pressure rapidly, causing nitrogen bubbles to form within the cells and expand, breaking the cell walls. This releases flavor compounds quickly, allowing them to easily dissolve in solution and permeate other ingredients. It even tenderizes meats.

The possibilities are really endless, but check out makezine.com/go/nitrogen-cavitation for some recipe suggestions.

BALSAMIC CAVIAR

1. Pour the olive oil into a tall glass and place the glass into the freezer for about 30 minutes (Figure Ⓐ).

2. Pour the balsamic vinegar into a saucepan and sprinkle with agar (Figures Ⓑ and Ⓒ). Whisk constantly and bring to a boil. Once boiling, remove the mixture from the heat and skim to remove any impurities.

3. When the temperature drops to between 120°F and 130°F (Figure Ⓓ), fill a syringe with the agar solution and carefully drip one drop at a time into the cold oil (Figure Ⓔ). The syringe should be high enough that the drops sink into the oil, but not so high that they split into smaller droplets. The spheres can be stored in the oil in the refrigerator until they are ready to be served.

BRUSCHETTA

1. Preheat the oven to 350°F.

2. Dice the red onion and place it into a bowl. Add the lemon juice and let sit for at least 15 minutes.

3. Slice the baguette diagonally. Place the slices onto a baking sheet and drizzle with olive oil and a pinch of salt. Bake until toasted, approximately 20 minutes.

4. Dice the tomatoes and mince the garlic. Mix both into the bowl of red onions.

5. Remove the toasted bread from the oven and place a slice of mozzarella on each.

6. Arrange two large basil leaves on each mozzarella slice and top with a spoonful of the tomato mixture.

7. Gently scoop the balsamic caviar from the oil and spoon it onto the tomato mixture. Add salt and pepper to taste, and serve immediately. ✏

Ⓐ

Ⓑ

Ⓒ

Ⓓ

Ⓔ

Doing Science with Drones

Written by Forrest M. Mims III

Aerial imaging and remote sensing have never been easier

FORREST M. MIMS III (forrestmims.org), an amateur scientist and Rolex Award winner, was named by *Discover* magazine as one of the "50 Best Brains in Science." His books have sold more than 7 million copies.

A A lightweight instrument platform can be taped together and then taped to a drone's landing gear.

**Time Required:
15–30 Minutes**

**Cost:
$5–$10**

MATERIALS

» **Drone with video camera**

» **Wood slats, ⅛"×1": 12" long (2) and 8" long (2)**

» **Duct tape**

» **Science instruments of your choice, lightweight enough to be carried by your drone.** I've flown a FLIR One thermal camera, an ultraviolet-B radiometer, and a data logger.

SMALL UNMANNED AERIAL VEHICLES (UAVS OR SIMPLY DRONES) HAVE BECOME AN IMPORTANT ADDITION to my science tool kit. Not only are they good for aerial imaging, they're ideal for measuring changes in temperature, relative humidity, dew point, and other parameters at various altitudes. The simplest way to use instruments that measure these parameters is to place your instrument's readout directly in the field of view of the drone's camera. This allows you to easily see changes in the data simply by viewing the scene on the drone's controller display. You can also photograph the readout with the camera of the DJI Phantom and other advanced drones to log the altitude, coordinates, and time directly into the photo's EXIF file.

SIMPLE INSTRUMENT PLATFORM

You can easily make a very lightweight platform for flying instruments from a Phantom 3, Phantom 4, or other drones, using two 8" and two 12" wood slats (⅛"×1") from a hobby shop. Attach the slats to one another with duct tape to form a rectangle, then attach the 12" sides to the

Forrest M. Mims III

GIVE A GIFT.
FULL YEAR ONLY $34.95.

GIFT FROM

NAME

ADDRESS

CITY   STATE   ZIP

COUNTRY

EMAIL ADDRESS

☐ Please send me my own full year subscription of Make: for $34.95.

GIFT TO

NAME

ADDRESS

CITY   STATE   ZIP

COUNTRY

EMAIL ADDRESS *required for access to the digital editions

We'll send a card announcing your gift. Your recipients can also choose to receive the digital edition at no extra cost.
Price is for U.S. only. For Canada, add $9 per subscription. For orders outside the U.S. and Canada, add $15.

47CGS1

BUSINESS REPLY MAIL
FIRST-CLASS MAIL   PERMIT NO. 865   NORTH HOLLYWOOD, CA

POSTAGE WILL BE PAID BY ADDRESSEE

Make:

PO BOX 17046
NORTH HOLLYWOOD CA 91615-9186

bottom of the drone's landing gear with duct tape. No metal should be used, to avoid possible interference with the drone's compass. The platform can carry a variety of instruments (Figure A).

TIP: While a drone may be able to fly properly when its center of gravity is changed, a counterweight opposite the payload will reduce the extra load on the motors required to stabilize the drone.

A FLYING ULTRAVIOLET RADIOMETER
The field where I have measured sunlight since 1990 is surrounded by trees. This doesn't affect measurements of direct sunlight, but the trees have begun to block some of the skylight measured by my full-sky radiometers. The question I've long asked is: How much sky blockage have the trees caused? A Phantom 3 drone has provided the answer.

The most important of my full sky measurements is the sun's ultraviolet-B radiation, and I have long used several DIY radiometers for this purpose. In 1995 and 1997 NASA assigned me to measure the ozone layer and UV-B in Brazil during the annual burning season when vast smoke plumes fill the sky. For these campaigns a UV-B radiometer that I built in 1994 was used, and that's the radiometer I wanted to fly atop a Phantom 3. This was a risky decision for such a venerable instrument, but I needed to learn the exact difference between the UV-B measured 5 feet and 70 feet over the ground.

The instrument platform described above was used for the test. The attachment magnets were removed from a 16-bit Onset data logger, which was then duct-taped to the platform (Figure B). The radiometer was very carefully taped to the exact center of the top of the Phantom (Figure C). The propeller tips were less than an inch away from the radiometer, so I made sure the radiometer and connection wires were firmly taped in place. Three flights answered the question: The trees were reducing the UV-B by 2.5%. After measuring the decline over the next 2–3 years, I can devise an algorithm to correct my past data back to 1994.

While sunlight instruments are best mounted at the highest point on a drone,

the Phantom 3's GPS system is also located there. Fortunately, the UV-B radiometer had no negative effect, perhaps because it's housed in a plastic enclosure and has relatively few metal parts.

CAUTION: If you mount an instrument atop your drone, be sure it is firmly attached and well away from propeller blades. Conduct an initial flight test near the ground to make sure the GPS and compass aren't adversely affected.

THERMAL AND VISIBLE IMAGERY OF NATURE
Thermal infrared cameras provide images of heat that reveal features that are completely invisible in standard visible light images. For example, thermal imagery can reveal subtle differences in soil moisture and texture, underground features, and even archaeological sites. Thermal images also reveal the presence of people and warm-blooded animals in daylight and at night.

Thermal cameras are available for drones, but they cost thousands of dollars. An inexpensive alternative is to use a thermal camera designed for a smartphone. These can be purchased for under $250. I've used a FLIR One thermal camera connected to an iPhone 5 to acquire thermal imagery from a Phantom 3. The camera-equipped phone was attached to the front of the instrument platform described above.

While this method of acquiring thermal images works, it's not perfect, for the camera is not stabilized like the drone's gimbal-mounted visible light camera. So, it's best to fly slowly on a still day. Another drawback is that the imagery must be acquired in movie mode since there is not a simple method to trigger the camera to acquire still shots. Nevertheless, the method does work, and it provides a useful tool to explore applications for thermal imagery before investing thousands of dollars in a drone with a gimbal-stabilized thermal camera.

For example, my rural place borders a city that recently installed a sewer system in a nearby subdivision. The main sewer pipe was placed in a deep, gravel-lined ditch that intercepted underground streams and created new springs, one of which emerged

An Onset 16-bit data logger taped to a drone instrument platform.

A UV-B radiometer taped atop a Phantom 3.

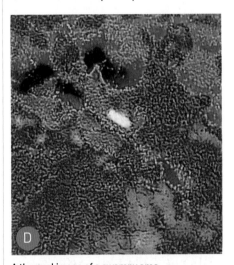
A thermal image of a swampy area.

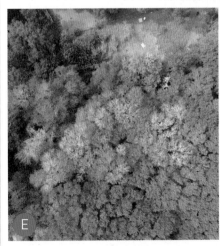
An aerial image of trees being killed by a new spring — more than a dozen had lost their leaves.

F A drone's-eye view of the swampy conditions under the canopy shown in Figure E.

by my driveway and another in the woods below our house. A thermal image from a Phantom 3 flown over the woods (Figure **D**, previous page) revealed cool standing water (black and dark blue), wet soil and grasses (light blue) and the warm tree canopy (red).

A Phantom 4 camera drone provided an ideal way to keep track of damage to trees caused by the new spring. Images from above the tree canopy clearly revealed that more than a dozen large trees had lost their leaves. (Figure **E**, previous page). Carefully flying the drone under the canopy provided much better images of the swamp beneath the trees (Figure **F**) than photos made from the ground.

SURVEYING AGRICULTURAL FIELDS

Farmers are using drones equipped with near-infrared cameras to detect diseases in their crops. Some DIYers have modified cameras to see near-IR by opening the

camera and removing its IR blocking filter, but even a standard visible wavelength camera can clearly show the extent of crop damage much better than a ground survey. While flying a Phantom 4 high over an 88-acre field of cotton, I noticed large dark areas amidst the white background of cotton being harvested (Figure **G**). The farmer told me that the dark areas were cotton that had been damaged by a fungus known as cotton root rot. While I didn't notice the diseased plants from the ground, they were clearly visible from the air.

MONITORING SCIENTIFIC INSTRUMENTS

Scientific instruments to monitor outdoor conditions are often placed in awkward locations. For example, the Department of Agriculture sunlight instruments I manage for Colorado State University are installed atop a two-story roof at Texas Lutheran

University. The alignment of the rotating shadow bands of three of these instruments must be checked weekly, which requires climbing a 12-foot steel ladder to a roof hatch. Should circumstances make the roof visit impossible, a drone can be used to check the instruments (Figure **H**). The drone's camera provides a very clear view of the three shadow bands (Figure **I**).

SAFETY PRECAUTIONS

The Federal Aviation Administration (FAA) has rules for flying drones, including flying no higher than 400 feet, not flying over people, and flying only from 30 minutes before sunrise to 30 minutes after sunset. Various rules apply when airports are within 5 miles. Drone pilots should register their UAS, fly with care, and learn more about FAA's UAS rules at www.faa.gov/uas and makezine.com/go/faa-drone-law. ◓

G A cotton field afflicted with the fungus known as cotton root rot (dark areas).

H Drone viewing rooftop instruments.

I Drone's view of the three rotating shadow bands.

Forrest M. Mims III

Hollow, Dolly!

Written by Bill Chellberg

BILL CHELLBERG was a printer for 50 years. He started collecting dolls when his first granddaughter was born, thinking she would surely play with them — wrong! But he was hooked and decided to try to make one himself.

**Time Required:
80 Hours
Cost:
$100**

MATERIALS

- » **No. 1 casting plaster** for moldmaking
- » **Mold release agent** I used green soap
- » **Sculpting medium** I used modeling clay
- » **Porcelain slip** available in many skin colors
- » **Porcelain paints** for lips, cheeks, and eyebrows
- » **Doll wig**
- » **Dolls eyes**
- » **Doll eyelashes**
- » **Eyewax**

TOOLS

- » **Sculpting tools**
- » **Sanding materials**
- » **Set of eye sizer tools**
- » **Eye positioner (optional)**
- » **Brushes**
- » **Material to make the mold sides** I used shelf boards with Formica-like surfaces.
- » **Adhesive for eyelashes**
- » **Clamps**
- » **Large knife** hunting or butcher

Build molds for creating custom porcelain collectibles

MAKERS MAKE THINGS FOR VARIOUS REASONS. MY REASON WAS THE CHALLENGE OF IT: I had just started collecting dolls and I wanted the most permanent (porcelain) and most child-like I could get.

It was a long process as I had never done any of it before, but deriving most of my help online, I learned how to sculpt, and how porcelain objects (Figure Ⓐ) become hollow (see box, right).

While I can't give instructions in sculpting — there are many good tutorials on YouTube — here is how I made forms for molding porcelain dolls from sculptures, focusing on the head. For additional information, visit this project online at makezine.com/go/DIY-porcelain-doll-molds.

How to Make Hollow Porcelain Objects

When the "slip" (a clay soup that you buy) is poured into the mold and allowed to set for a few minutes, the water in it begins to soak into the plaster mold and the clay in the slip hardens against the mold walls. Then the remaining slip is poured out.

Bill Chellberg

BUILD THE MOLD FORM

Construct two "L" shaped forms at least twice the size of the head sculpt, which will lay on its back in the mold. These forms should be no less than 3" higher than your sculpt from tip of the nose to back of the head. Clamp the "L" shapes into a box.

CREATE THE MOLD

1. Carefully suspend your original sculpt on a bed of clay in the bottom of the box, face up, so that it clears all sides (top, bottom, right, left, up, down) by about 1". Fill in clay so that it forms a tight seal around the lower half of the head (Figure B). Any protrusions above this middle line will not slip out of the mold later. I added ear plugs, which made this (in the end) a four-part mold. When the plugs are removed the other halves of the mold will open without damage to the ears.

2. Using a brush, apply a mold release agent — I used green soap — both on your sculpt and on the clay. Don't let it pool anywhere. Place 4 marbles halfway into the clay around the head to form register points for aligning the final molds.

3. Prepare plaster to the consistency of soft ice cream and slowly pour on the face of your sculpt, up to about 1" above the nose, tapping the forms occasionally to remove any air bubbles. Let dry overnight.

4. Remove the clamps and tap the forms as you gently remove them. Turn it plaster-side down and begin gently removing the clay from under the head. DO NOT remove the head from the plaster. Remove the marbles. Begin the molding process from Step 1 again, and pour another batch of plaster on the back of the head. Remove the mold forms after the plaster sets (Figure C).

OPEN THE MOLD

Place a large knife on the line where the 2 batches of molding plaster came together. With a small hammer, tap gently, increasing force until the two halves separate. Remove and store your original handiwork.

Next, cut the sprue opening — a hole tapered from large on the outside to smaller on the inside — in which to pour the slip (Figure D). This is usually done at the neck opening and is used to mount the final head to the shoulder plate.

USE THE MOLD
POUR THE SLIP

Align the two halves and bind them together – tight! Pour the slip steadily and slowly so as not to generate air bubbles. Over a catch pan, fill the sprue opening to the top, refilling as the slip sinks. When a coating roughly the thickness of a nickel forms on the inside of the sprue (about 10 minutes), slowly pour the remaining slip back into its original container. Let the mold set for up to four hours — I place mine over the original slip container to continue draining.

> **WARNING:** Don't pour slip down a drain or toilet — it will clog!

RELEASE MOLD

Gently open the mold. The greenware will be very soft, like leather — be careful, it can collapse! If you can handle it at this point, cut the large circular opening, called the pate opening (Figure E), and set the head on that opening to dry for 24 hours.

CUT EYE OPENINGS

Sand — very gently — and do light remodeling if you'd like. Cut the openings to accommodate the eyes (glass or plastic) (Figure F), then use the eye sizers from the back of the head to smooth and set them to the proper depth.

FIRE AND FINISH

Soft-fire the head and smooth using diamond sanding pads. Apply paint, high-fire, then check that the eyes fit. If you are unhappy with the colors, go over them and re-fire. Position the eyes using eye wax, which allows them to be turned and focused, then plaster over the back to hold them in place.

> **TIP:** If you do not have a kiln, check with schools or doll stores.

You can now add the hair, eyelashes and other finishing touches. Looking for ideas or have questions? Doll shows are excellent sources for supplies and help. ◯

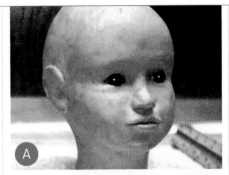
A

B

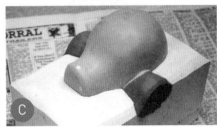
C

D

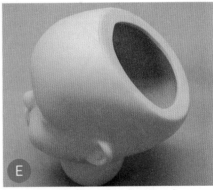
E

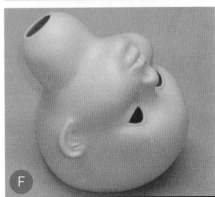
F

For more details and to share your doll-making projects, visit this project online at makezine.com/go/DIY-porcelain-doll-molds.

Toy Inventor's Notebook
WHIRLY NOISEMAKER
Invented and drawn by Bob Knetzger

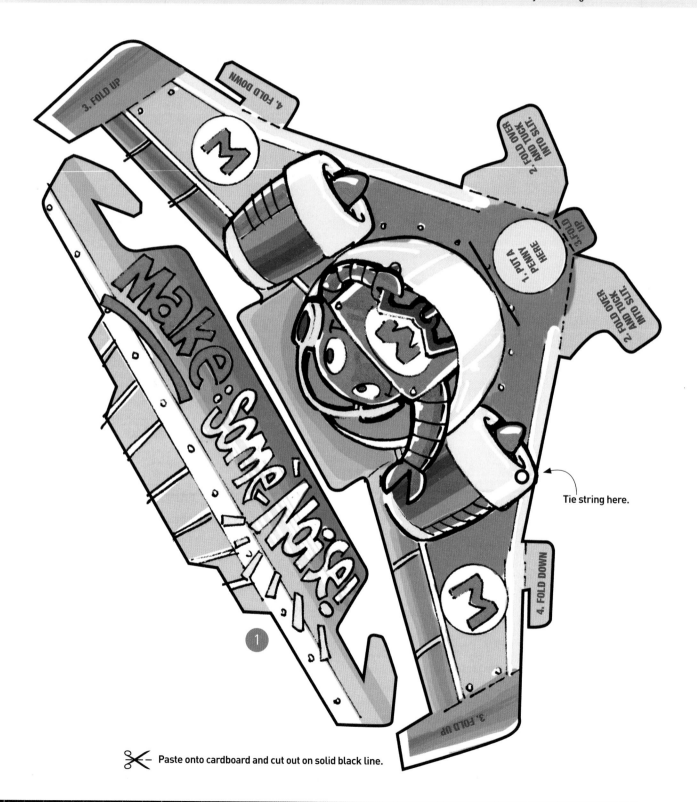

Tie string here.

✂ Paste onto cardboard and cut out on solid black line.

Whizz it around your head and hear it roar!

Time Required: 15-30 Minutes
Cost: $1-$2

MATERIALS
» **Thin cardboard like from a cereal box**
» **String or monofilament fishing line**
» **Paste or glue**
» **A penny**

TOOLS
» **Scissors**
» **Hole punch**

HERE'S AN UPDATE OF A BULLROARER, AN ANCIENT NOISEMAKER DATING TO THE PALEOLITHIC PERIOD and found in many cultures around the world. When a thin, wooden paddle is whirled around on a string it creates a roaring sound with a cool 360° "surround sound" effect. This version is based on a mid-20th-century cardboard toy premium and makes a fun flapping sound: the faster you whirl it around, the louder and higher the buzz!

1 **Tear out the opposite page** and paste it onto a piece of thin cardboard. Then cut out the noisemaker on the solid black line. Be sure to also cut the slit in the center.

You can also go online to makezine.com/go/ whirly-noisemaker to download and print out this Makey bot version, or a blank version to finish with your own graphics.

2 **Assemble:** To make a nose weight, place a penny where indicated and fold the tabs over and into the slit. Bend the nose tab and wing tips *up* and the flapper fingers *down*. Turn the plane over and hook the flapper onto the fingers. The plane body will be slightly curved. Punch a small hole at the edge of the wing and tie on a string or monofilament.

3 **Whirl the noisemaker around** and around over your head. Start slowly to get the flapper going and then speed up and down to change the sound. ◗

Download the template and share your toys at makezine.com/go/whirly-noisemaker.

WARNING: The flapping sound and whirling action make this an irresistible cat toy!

Caleb Kraft

BROTHER SE400
SEWING AND EMBROIDERY MACHINE $400 brother-usa.com

This combination sewing and embroidery machine is relatively affordable and packed with features. It's computerized, so you can drop in a design and the machine will physically move a gantry to embroider your pattern into the fabric.

I picked it up to create some patches, but the computerization isn't limited to embroidery — there are also 67 built-in stitches at the press of a button. Another aspect I love is the auto-threading system —

needles are tiny, so threading them with the push of a lever is glorious.

Embroidery is extremely easy: You load your design, load your thread and bobbin, and hit a button. It will stitch a color, then pause so you can switch to the next color before you press the button to proceed. The machine is fairly quiet, and also has a level of intelligence: It will stop if it runs out of thread or something else goes wrong so that you can correct the issue.

You'll have to find third party software for designing embroidery, and you'll want to keep the manual handy, as some of the interface icons are nonintuitive. I also had to discover on my own that one of my designs wasn't working because it was larger than the 4"×4" embroidery hoop — there was no error message to guide me. Otherwise, I've really enjoyed playing with this machine and I've got a long list of ideas I'm looking forward to creating. —*Caleb Kraft*

BOSCH 12V IMPACT DRIVER

$169 boschtools.com

I was able to use Bosch's small 12V tools a few years ago on some projects for a TV show, and quickly discovered they packed more than enough power for almost anything I was working on without the muscle-burning bulk of large 18–20V alternatives. I'm now using the latest model impact driver around my house — it's even smaller and stronger than before. I have to be careful with it though; it was great for building a small skate ramp, but I've stripped a few screws assembling furniture with it.

—*Mike Senese*

MILWAUKEE M18 CORDLESS 10" MITER SAW

$599 milwaukeetool.com

This glorious beast of a sliding compound bevel miter saw is missing one thing: a power cord. That's right, it's battery powered. I never thought I'd see the day when a full-powered, 10" power saw could get away with this, but this one does so with gusto. The M18 Fuel 9.0Ah battery pack gives it the juice it needs to go and go — up to 300 chops through 2×4s, according to Milwaukee.

I've been packing it in the back of my Jeep and taking it to a friend's house to build some structures. It works great. The 45lb tool folds up and locks the slide and table nicely, and has handles to lug it. The convenience of being able to make cuts anywhere you want without the need for a power extension is incredible. —*John Edgar Park*

STRONG HAND ADJUSTABLE MAGNETIC V-PADS

$27 stronghandtools.com

The more I make, the more time I spend on "workholding," which is the use of jigs, fixtures, and tools to secure a piece while cutting, grinding, painting, welding, etc. I often find that I'll spend 10 minutes setting up for a 20 second tack weld. It may sound tedious, but it's become something I really enjoy. Nevertheless, I'm always eager to discover a product that speeds up or improves my workholding. I've started using Strong Hand's Magnetic V-Pads. These have two magnets that swivel 90° on the ends of a bracket. This allows the magnets to hold things in a startling variety of configurations. You can hold plates flush, stand a plate on edge against a pipe, clamp tubes at 45° to a plate, and thousands of other configurations. I've used magnetic angle clamps before, but these clamps' ability to change position makes them wildly more useful and they're fast to set up. V-Pads come in a variety of sizes from a number of different vendors, but this pack from Strong Hand Tools is a good start, with two large and two regular pads and a holding force of 18lbs. If I lost or mangled the ones I have, I'd instantly replace them. —*Tim Deagan*

IFIXIT ESSENTIAL ELECTRONICS TOOLKIT V2.0

$20 ifixit.com

Like other iFixit tool kits, the new Essential Electronics Toolkit is focused on consumer electronics repair tasks, such as swapping out defective smartphone parts. I rarely need to fix my electronic devices (knock on wood), but iFixit tools are also quite capable for use in general electronics projects.

This $20 kit gives you a very good selection of screwdriver bits, prying and lifting tools, and a nice magnetically latched storage case with a sorting tray lid. It also comes with a suction cup handle, designed specifically for opening electronics products.

I use spudgers for soldering tasks, the prying tool as a chip lifter and device opener, and opening picks as precision scrapers. The screwdriver is good for everyday use, and the tweezers I keep as a backup. —*Stuart Deutsch*

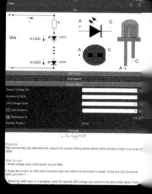

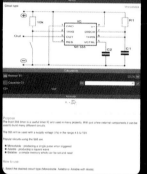

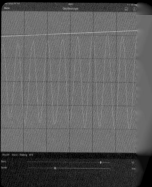

SERVOCITY DIGITAL MANUAL SPEED CONTROLLER

$80 servocity.com

ServoCity's digital manual speed controller is a quick and easy way to test and control gearmotors without needing to do any programming. It's fairly compact (2.4"× 2.4"× 0.875") and has a polished appearance. There are mounting holes on the bottom, but double-sided tape also works well to fix the controller in place.

This is a good option for use with finished projects, such as a motorized slider or turntable, or for prototyping your next robotics project.

It features bidirectional speed control, can deliver up to 10A of continuous current, and can be powered by a 6–16V DC power source. There's a built-in Tamiya-style battery connector which you can cut off to attach your own connector, and a 2.5mm barrel jack. It's a nice and simple controller that gets your motor spinning with minimal effort. —*SD*

ELECTRONIC TOOLBOX PRO

$7 iTunes store; $6 Windows store electronic-toolb

The Electronic Toolbox is a technically advanced, all-in-one app for anyone working in electronics. It features over 77 in and more than 150 applications, including a PCB trace tool, tools, extensive component reference guides, a frequency g and even an oscilloscope. The app also supports Dropbox a storing options as well as in-app storage. The interface is s to use functionality and a fully customizable experience. It c improve with constant development. This is simply a must f engineers, or anyone working intimately with electronics. —

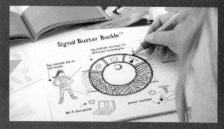

THE EXTRAORDINAIRES DESIGN STUDIO PR

$80 creativityhub.com

One of the many wonderful aspects of the maker movement is getting kids (and involved in thinking about *how* to make things. But teaching design and enginee can be difficult, especially to kids. *The Extraordinaires Design Studio Pro* is a desi disguised as a game, which cleverly teaches user-centered design, and outside thinking to teens. The game focuses on a series of characters, the Extraordinair pressing design needs. The Extraordinaires include the Evil Genius, an Alien, a Teen Vampire, and a steampunk Time Traveler. The two-sided cards for each of clients are beautifully rendered, and offer clues to the lives of the characters, th design constraints. The "Think Cards" help you reason through a design, and wh ready, you commit that design to the Idea Pad and present it to the group. Desig given at the end. A 120-page guidebook provides a basic crash course in playing and in user-centered design. —*Gareth Branwyn*

FISHER-PRICE THINK & LEARN CODE-A-PILLAR
$50 fisher-price.mattel.com

CUBETTO
$225 primotoys.com

These two little robots take very different approaches, but both are hits with my young kids, who have no idea they're learning computer programming concepts. And it has an added bonus — no screen time is involved.

Code-A-Pillar is a Fisher-Price toy, which means all-American style: it's cartoon-cute, plastic, and *loud*, with lots of flashing lights, music, and goofy sounds. The hardware itself is the programming interface — each segment of the caterpillar is marked with its function (go straight, turn left, play music, etc.) and you rearrange them as you please. Press the button and the robot drives off to execute your sequence. Additional segments (including new turn angles and sounds, and a repeat function) are offered, but we haven't tried those yet.

Cubetto, funded by a huge Kickstarter campaign last year, has a much calmer aesthetic, very European and Montessori-approved. The robot is a demure wooden cube

that emits only quiet bleeps. On a separate, Bluetooth-paired board, you sequence Cubetto's movement by placing colored blocks (forward, right, left, or "function") along a flow-path. My older child, 5½, quickly became fascinated with this literal drag-and-drop programming, and there's a little side loop (invoked by the function block) that teaches the concept of subroutines. Cubetto moves much shorter distances than Code-A-Pillar, making it suitable for smaller play areas; it comes with a play map and storybook with sample missions to try.

In both cases, it's challenging and fun for kids to program their little robot without screens or language — fun enough that they'll endure, and even enjoy, the inevitable trial-and-error. When at last the robot miraculously navigates through the dining-chair forest, between the sibling's feet, and coasts to a stop without a collision — the victory cheer is worth all the effort.

—*Keith Hammond*

SAM LABS CURIOUS CARS KIT
$200 samlabs.com

This kit is a fun, simple combination of Bluetooth widgets and pop-and-fold papercraft cars. The modules, which include motors, an LED, slider, and tilt sensor, can be easily connected, programmed, and controlled via the Curious Cars app. You can follow the adorable in-app storyline, which uses a maze-like interface to drag and connect the different modules. It's a bit tricky for those with wider fingers, but little makers will have no trouble. Be sure not to close the app halfway through constructing a car, or prepare to re-do your mazes. When you're ready, dive into the project ideas booklet for more ambitious applications (some call for additional items like glue or a plastic bowl). —*Sophia Smith*

GLOWFORGE

Ease of use and ingenious features make this machine worth the wait *Written by Matt Stultz*

IN MAY OF 2015 DAN SHAPIRO TOOK THE STAGE OF MAKERCON BAY AREA
to announce he was starting a company to manufacture a groundbreaking laser cutter, the Glowforge. I talked to Shapiro and we agreed that when they were getting close to shipment, he would send me a review unit. Preorder customers (including my wife) know the official launch has been fraught with delays, but a few weeks ago mine arrived. Let me say, it is worth the wait!

UP AND CUTTING IN 10 MINUTES
Unpacking was quick and easy. Following the instructions, I had it plugged in, set up, and was making my first cut within 10 minutes. The only data connection to the machine is via wi-fi, which is simple to set up. There is no software to install — instead a web app on the Glowforge website allows users to control the machine and ensure software is always up to date.

The software will accept a number of image files for etching, but like most laser cutters, the mainstay is vector graphics — in the Glowforge's case, SVG files. I'm an Inkscape fan, and while many other digital fabrication platforms stumble on Inkscape-created SVG files, the Glowforge handles them perfectly.

BETTER THROUGH BETA
My first cuts were clean and easy. The machine performed exactly as I would expect. For the most part laser cutters either work or they don't. The real trick is in how easy they are to operate. The built-in camera and the software work together to make the Glowforge the easiest digital fabrication machine I have used.

Of course, this machine is still beta and there are a few problems. I created a test piece to make it easy to examine all the modes of the Glowforge (Cut, 3 levels of Engrave, and 2 levels of Score). The engrave sections show points where the laser didn't start or complete lines at the correct position. The score lines have dots on each end, again showing timing issues caused by the laser coming on before movement had begun. The good news is that these are all software problems that the team can fix while they are still building your unit.

HANG IN THERE
While I know a lot of customers and the community at large have lost some confidence in the Glowforge shipping, I would suggest holding out. It looks like the wait is almost over, and there is an amazing machine on its way. ⏻

- **MANUFACTURER** Glowforge
- **PRICE AS TESTED** $2,995
- **CUTTING AREA** 290mm×515mm
- **CUT UNTETHERED?** Yes (over wi-fi)
- **ONBOARD CONTROLS?** Yes (single control button)
- **CONTROL SOFTWARE** Glowforge web interface, no installed software
- **OS** Windows, Mac, Linux
- **OPEN SOFTWARE?** No
- **OPEN HARDWARE?** No

glowforge.com

PRO TIPS
Use color mapping on your SVG files to change the order and type of operation on parts of your design.

Make sure to follow the guidelines for venting — the built-in fan can do its job but not if you make it work too hard.

WHY TO BUY
The Glowforge is an extremely easy-to-use laser cutter that takes away many of the software pain points that plague other machines.

MATT STULTZ is the 3D printing and digital fabrication lead for *Make:*. He is also the founder and organizer of 3DPPVD and Ocean State Maker Mill, where he spends his time tinkering in Rhode Island.

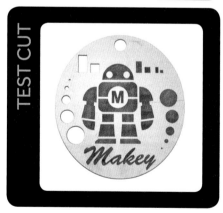

TEST CUT

Matt Stultz

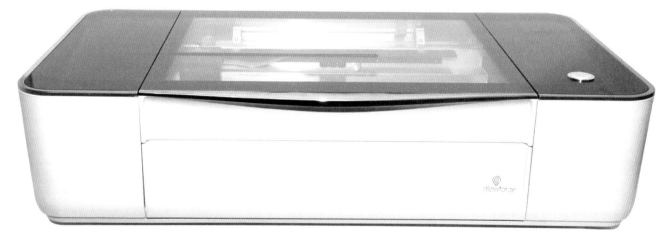

Make: Essential Reads

BOOKS

GETTING STARTED WITH RASPBERRY PI, THIRD EDITION
By Matt Richardson + Shawn Wallace $20

Updated to include coverage of the Raspberry Pi Models 2 and 3, this book takes you step-by-step through many fun and educational possibilities. If you're learning how to program — or looking to build new electronic projects — this hands-on guide will show you just how valuable that flexible little platform can be.

GETTING STARTED WITH ARDUINO, THIRD EDITION
By Massimo Banzi + Michael Shiloh $20

This thorough introduction, updated for the latest Arduino release, helps you start prototyping right away. From obtaining the required components to putting the final touches on your project, all the information you need is here!

GETTING STARTED WITH SENSORS
By Kimmo Karvinen + Tero Karvinen $20

This book starts by teaching you the basic electronic circuits to read and react to sensor data. It then goes on to show how to use Arduino to develop sensor systems, plus build sensor projects with the Linux-powered Raspberry Pi.

MAKE: ACTION
By Simon Monk $35

This book clearly explains the differences between Arduino and Raspberry Pi, when to use them, and which purposes best suit each platform. You'll learn to control LEDs, motors of various types, solenoids, AC devices, heaters, coolers, displays, and sound — as well as commanding these devices over the internet.

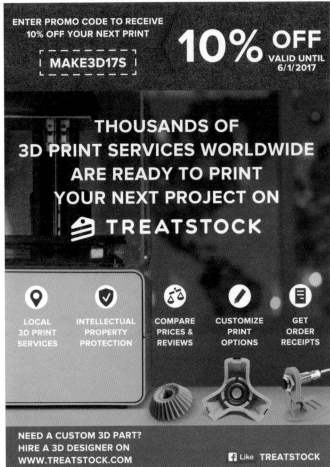

SHOW&TELL

Dazzling projects from inventive makers like you

Sharing what you've made is half the joy of making! Check out these makers on Instagram, and show us your photos by tagging @makemagazine.

1. Reminiscent of a group of barnacles, this textural ceramic by **Courtney Diedrich** (@c.diedrich) is made up of hundreds of individual parts that are joined together using slip to create larger, more complex forms.

2. **Charles Staffeld** (@newcharleslikesmovies) designed this PCB for the driver section of a Class B audio amplifier. It uses a bunch of small signal transistors in parallel to put upwards of 20W–30W into a 4-ohm load.

3. Woodworker **Steve Hadeka** of Pleasant Ranch (pleasantranch.com) tends to sell birdhouses and bottle openers, but decided to turn this beautiful black walnut

4. A self-described silversmith, artist, and constant crafter, **English Norman** (@englishnorman) made these metallic earrings inspired by kinetic sculpture.

5. This was **Morgan Lembke**'s (@mlembke98) first project! She crafted this lovely thread art inspired by Pinterest and her love for her home state.

6. Metalworker **Harrison Tucker** (@odin_craft) hand carves solid brass goods, including skull knuckles, lanyard beads, playing cards, and rings.

7. This wither strap with oak leaf detail is by **Kelly Devins**

leatherworker who does everything by hand, from the cutting of the leather to the tooling, stitching, and staining.

8. Tinkerer **Gustavo Tuntisi** (@1of1tocovet) is hard at work on this Nixie clock, with a large IN14 Nixie tube, an OG4 dekatron tube, and a logic output connected to a chime that sounds off every hour on the hour.

9. This work in progress, built on veroboard, comes from **Thomas Cassidy** (@tomcassidywasps), who's trying to improve an old schematic called "The Gristleizer."